Generalized Hypergeometric Functions

Transformations and group theoretical aspects

Generalized Hypergeometric Functions

Transformations and group theoretical aspects

K Srinivasa Rao
Institute of Mathematical Sciences, Madras (Chennai), India

Vasudevan Lakshminarayanan
University of Waterloo, Ontario, Canada

IOP Publishing, Bristol, UK

ISBN 978-0-7503-1496-1 (ebook)
ISBN 978-0-7503-1494-7 (print)
ISBN 978-0-7503-1495-4 (mobi)

DOI 10.1088/978-0-7503-1496-1

Version: 20181001

IOP Expanding Physics
ISSN 2053-2563 (online)
ISSN 2054-7315 (print)

British Library Cataloguing-in-Publication Data: A catalogue record for this book is available from the British Library.

Published by IOP Publishing, wholly owned by The Institute of Physics, London

IOP Publishing, Temple Circus, Temple Way, Bristol, BS1 6HG, UK

US Office: IOP Publishing, Inc., 190 North Independence Mall West, Suite 601, Philadelphia, PA 19106, USA

Dr Geetha Srinivasa Rao, FNASc (8 May 1944–24 July 2016)
Professor, Ramanujan Institute for Advanced Study in Mathematics, University of Madras, Chennai and President, Indian Mathematical Society, 2012 (KSR)

Nan, la stella della mia vita e la mia compagna di crimine (VL)

Contents

Preface

My work has always tried to unite the True with the Beautiful and when I had to choose one or the other, I usually chose the Beautiful.

Hermann Weyl

(as quoted by Freeman Dysan in Obituary for Herman Weyl, Nature, 177:458 (1956))

Beauty is Truth, and Truth Beauty,
That is all ye know, and all ye need to know.

John Keats

(Ode on a Grecian urn)

In this monograph, we consider some group theoretical aspects of the Gauss hypergeometric function and their transformations. The intimate connection between hypergeometric functions and the special functions of mathematics has been stated succinctly as a theorem by W W Bell (1968). The relationship between the special functions and their corresponding hypergeometric functions is given in block diagram form by Lakshminarayanan and Varadarajan (2015, figure 21.1). This latter book is a prequel to the work presented here. Along with the book by Srinivasa Rao and Rajeswari (1993), the present monograph forms a trilogy.

This monograph has nine chapters. In chapter 1, we present the hypergeometric series introduced in 1812 by Gauss, along with the hypergeometric function as a function of four variables: the differential equation satisfied by the function, its beta integral representation (due to Euler), and the special functions associated with it. The 24 solutions of the second order differential equation of Gauss, due to Kummmer, are given in an appendix to this chapter.

In chapter 2, an overview of the basic concepts and definitions in group theory, necessary to understand the group theoretical aspects of hypergeometric functions is presented. A brief introduction to Lie groups and Lie algebras is provided, exemplified by a typical group of fundamental importance and relevance in physics; namely, the rotation group. The ubiquitous symmetric group S_n and a new interesting property of S_n, extracted by assigning to its elements unique place values, is also presented.

In chapter 3, the group theory of the Kummer solutions of the Gauss differential equation is presented and it reveals the power of group theory in unification and simplification. The 24 solutions of Kummer are shown to be in one-to-one correspondence with the 24 symmetries of an ordinary cube—essentially a map of the four variables of the Gauss hypergeometric function a, b, c, z onto the six variables x_1, x_2, x_3, x_4, x_5, x_6 associated with the six sides of a cube. The relationship of the Gauss second order ordinary differential equation to the Riemann equation is pointed out.

In chapter 4, after introducing the generalized hypergeometric functions with $p + 1$ numerator variables and p denominator variables, the Whipple notation for the ${}_3F_2(1)$ is introduced. A recursive use of the Weber–Erdélyi transformation is shown to result in a group of 72 transformations for a terminating ${}_3F_2(1)$ series, which constitutes a new group G_T. The group theoretical structure of G_T, and its invariant subgroups are presented. The structure is more intricate than that of the symmetric group S_5, which is shown to be the invariance group for the non-terminating ${}_3F_2(1)$ series.

In chapter 5, the origins of the quantum theory of angular momentum and the algebra associated with it—the two- and three-dimensional representations of the angular momentum operators, the rotation group, the Thomae transformation, which relates the different forms of the ${}_3F_2(1)$ for the 3-j coefficient and the two equivalent sets of three and four ${}_4F_3(1)$s necessary and sufficient to account for the 144 symmetries of the 6-j coefficient—are presented.

The concept of the recoupling coefficient introduced by Guilio Racah for three angular momenta is introduced in chapter 6. This Racah—or, 6-j—coefficient gives rise to the Racah–Wigner algebra essential today in the study of complex (atomic, nuclear, and molecular) spectra. The Regge symmetries of the Racah coefficient result in 144 symmetries of the 6-j coefficient. Relating the 6-j coefficient to a hypergeometric function of unit argument results in two equivalent sets of ${}_4F_3(1)$, which are shown to be necessary and sufficient to account for the 144 symmetries of the coefficient. These sets are shown to be related to each other through the reversal of the ${}_4F_3(1)$ series. A recursive use of the Bailey transformation results in a set of transformations that form a group. The group theory of these transformations is then presented in this chapter.

In chapter 7, a generalization of the Gauss series in one variable—due to Kummer are given here in the Appendix, to double the hypergeometric series—is presented. The generalization related to the products of ${}_pF_q$s give rise to a double and triple hypergeometric series of unit arguments. The 9-j or $\ell s - jj$ coefficient of the angular momentum is defined and identified to be a special case of the triple hypergeometric series. This identification enables the development of a new algorithm for the numerical computation of the 9-j coefficient; to define polynomial zeros of this coefficient and lead to the discovery of new transformation formulas involving the 9-j coefficient. The concept of the stretched 9-j coefficient is introduced and the summation and transformation formulas are derived for the multiple hypergeometric series.

In chapter 8, a new method is proposed based on the evaluation of beta integrals to derive new hypergeometric identities from old ones. We call this the *beta integral method.* We show that we can systematically generate new hypergeometric transformations from older existing transformations by this beta integral method.

In chapter 9, we present in brief the work on the hypergeometric series by Gauss, its further developments in Europe, and the genius of the Indian mathematician Srinivasa Ramanujan, which enabled him to discover the results for himself with only a hint of the Gauss summation theorem in Carr's 1886 book, *A Synopsis of Elementary Results in Pure Mathematics*; containing propositions, formulae, and methods of analysis with abridged demonstrations but without proofs (Carr 2002). With that as a foundation, Ramanujan discovered and recorded about 4000 entries in three notebooks—now renowned in the world of mathematics as *Ramanujan's Notebooks*—in a short life span of 32 years, 4 months and 4 days. We also present a new theorem, which arises from Ramanujan's entries in his notebooks (Srinivasa Rao *et al* 2005).

References

Bell W W 1968 *Special Functions for Scientists and Engineers* (New York: von Nostrand)

Carr G S 2002 *A Synopsis of Results in Pure and Applied Mathematics* Vol 2 (Cambridge: Cambridge University Press) Ist reprint edition 2002

Lakshminarayanan V and Varadharajan L S 2015 *Special Functions for Optical Science and Engineering*Vol 2 (Bellingham, WA: SPIE Press)

Srinivasa Rao K and Rajeswari V 1993 *Selected Topics in Quantum Theory of Angular Momentum* (New York: Springer)

Srinivasa Rao K, Vanden Berghe G and Krattenthaler C 2005 *J. Comput. Appl. Math.* **173** 239

Acknowledgements

KSR would like to thank Mr Ramesh Subramanian and the Indian Institute of Technology, Tada, Andhra Pradesh, India for giving him the opportunity to offer a special course in science (including group theory) to undergraduate students (2014–16). The EEC project at the University of Ghent, Belgium, carried out in collaboration with Professors Guido Venden Berghe and Joris Vander Jeugt as well as the visit to the University of Vienna at the invitation of Professor Christian Krattenthaler, is gratefully acknowledged. KSR also thanks Ms Usha Satyamurthy, Library Assistant at the Institute for Mathematical Sciences, Chennai for providing essential reference material whenever requested. He is also thankful to the families of his sons, Dr K S Aravind and Mr K S Anand for their hospitality during his stay at their homes in the last quarter of 2017, especially during the completion of the final stages of the manuscript for providing a peaceful, serene home environment in Portland, Oregon.

VL would like to thank Mr Peyman Gholami for help with the final LaTeX files. He also acknowledges the support of his research by various Discovery and Engage grants from the Natural Sciences and Engineering Research Council of Canada. He also thanks Nan, Bridie, Kelly, and Molly for their love and support in all his intellectual (and other) pursuits.

Both of us would like to thank Ms Ashley Gasque, the acquisitions editor at the Institute of Physics Publishing for her enthusiastic support of this project.

Author biographies

K Srinivasa Rao

KSR trained in theoretical nuclear physics initially and is now retired as a Senior Professor of the Institute of Mathematical Sciences, Madras (Chennai) and the Director (Hon) of the Srinivasa Ramanujan Academy of Maths Talent, Chennai. He was a Distinguished-DST Ramanujan Professor for Mathematical Sciences, (2005–2009), at the Srinivasa Ramanujan Center of the SASTRA University, Kumbakonam. In addition to nuclear physics research he has worked extensively on generalized hypergeometric functions, and pioneered the use of group theoretical methods on such functions and their transformations. His many honors include an Alexander von Humboldt Stitfung fellowship, Fellow of the Indian National Science Academy, the Tamilnadu Scientist Award, and various awards for popularization of Science from the Tamilnadu State Council for Science and Technology, the Srinivasa Ramanujan Academy of Maths Talent, Chennai, and the Ramanujan Mathematics Academy and Mathematics Library, Ramachandrapuram, Andhra Pradesh. His tenure as a Humboldt Fellow was spent at the Institute of Theoretical Nuclear Physics, University of Bonn, and at the Arnold Sommerfeld Center for Theoretical Physics of the Technical University of Clausthal, Germany. He was a visiting professor at the Rennselaer Polytechnic Institute, (1978–79), McGill University, McMaster University, Catholic University of America, Duke, University of Texas at Austin, and UNAM, Mexico City. As a recipient of an INSA-Royal Society, London, exchange scholarship, he has visited the Universities of Cambridge, London, Oxford, and Southampton in the UK. He served as the PI of an Indo-Belgian collaborative Project (supported by the EEC) on Quantum Theory of Angular Momentum, at the University of Ghent, Belgium for four years. Additionally, he was the PI on a multi-media project by the National Multimedia Resource Center of the Center for Development of Advance Computing, at Pune. This work has been hailed as a contribution to mathematics and mathematicians. This resulted in a two CD-ROM on the Life and Work of Srinivasa Ramanujan, an Indian Mathematical Genius. Other outreach activities include exhibits on the life and work of Srinivasa Ramanujan for the Indian Science Congress Exhibition (1999), ISCE (2012), the Visveswaraiah Museum, Bangalore, (2010), and at the Indian Institute for Science Education Research, Sri City (2018), both in Andhra Pradesh. He has published 219 articles and 8 books.

Vasudevan Lakshminarayanan

VL is currently at University of Waterloo, where he is a professor of vision science, physics, electrical and computer engineering,and systems design engineering. He was a KITP Scholar at the Kavli Institute for Theoretical Physics at UC Santa Barbara, an associate of the Michigan Center for Theoretical Physics and has held research and teaching positions at UC Irvine, UC Berkeley, University of Michigan, and the University of Missouri amongst others. He is also an adjunct professor of Electrical and Computer Engineering at Ryerson University, Toronto. He has been a visiting professor of Physics at IIT Delhi, a Royal Society of Edinburgh visiting professor at Glasgow, an international visiting professor of biophotonics in the Department of Information Engineering at University Degli Studii Brescia, Italy, and a Gian Professor of Physics at IIT Madras. He also served on UNESCO's International Year of Light planning committee, is a founding member of the UNESCO ALOP Program and is on the committee for the International Day of Light. He is on the optics advisory board of the International Center for Theoretical Physics at Trieste, Italy (since 2003), a consultant to the medical devices group of the US FDA (since 2011), he has represented the United States at two IUPAP general assemblies, was the chair of the US advisory committee for the International Commission on Optics, chair of the committee on international scientific affairs of the APS, member of the Public Policy Committee of APS (since 2017), a AAAS Science and technology policy fellow finalist, and a director of the OSA. He is a fellow of the American Physical Society, AAAS, Optical Society of America, SPIE–the International Society for Optics and Optical Engineering, Institute of Physics, and other professional bodies. He has worked in a number of areas ranging from quantum physics and spectroscopy, to bioengineering, mathematical optics, optical physics and engineering, image processing, optometry, ophthalmology, applied mathematics, science policy and cognitive neuroscience. He has over 350 publications and is the editor/coauthor of about 20 books, (most recently *Special Functions for Optical Science and Engineering* (SPIE Press, 2015) and *Understanding Optics with Python* (CRC Press, 2018)), and is the recipient of a number of Awards including the SPIE Optics educator award (2011) and the Esther Beller Hoffman medal of OSA (2013). He has been, or is, the technical editor/associate editor for a number of journals, including *Optics Letters* and *Journal of Modern Optics*, and serves on many NIH study sections (including those on bioengineering, neurotechologies, etc) as well as other agencies.

Chapter 1

Hypergeometric series

1.1 Introduction

The 'geometric' series:

$$1 + x + x^2 + x^3 + \ldots = (1 - x)^{-1} \tag{1.1}$$

is so called because the second term is referred to as the linear term, the third term is called the 'square' term, and the fourth the 'cube' term, since they refer to the area of a square and the volume of a cube, respectively. The fourth term onwards is called the 'hyper' cube terms—'hyper' from the Greek root meaning 'above' or 'beyond'. Equation (1.1) is also a binomial series and this equation may be interpreted as a summation theorem. This is because it represents the sum of the infinite geometric series as $(1 - x)^{-1}$, $\forall\ |x| < 1$.

John Wallis (1616–1703), in his book *Arithmatica Infinitorium* (1655), extended the ordinary geometric series to the hypergeometric series:

$$1 + a + a(a + b) + a(a + b)(a + 2b) + \ldots, \tag{1.2}$$

with the nth term given by:

$$a(a + b)(a + 2b)\cdots(a + (n - 1)b), \tag{1.3}$$

which later became, with $b = 1$, the Pochammer symbol:

$$(a)_n = a(a + 1)(a + 2)\ldots(a + n - 1) = \prod_{k=1}^{n} (a + k - 1), \tag{1.4}$$

where n is a non-negative integer named after Leo August Pochhammer (1841–1920), who used this notation for the rising factorial—sometimes also called the Pochhammer—function, Pochhammer polynomial, ascending factorial, rising sequential product, or the upper factorial.

Similar series' were studied by many other mathematicians, notably Leonhard Euler (1707–83) and Alexandre-Theophile Vandermonde (1735–96). Euler introduced the power series expansion of the form:

$$1 + \frac{ab}{c}\frac{z}{1!} + \frac{a(a+1)b(b+1)}{c(c+1)}\frac{z^2}{2!} + \frac{a(a+1)(a+2)b(b+1)(b+2)}{c(c+1)(c+2)}\frac{z^3}{3!} + \dots \\ = F(a, b, c, z), \tag{1.5}$$

where a, b, c are rational functions and $F(a, b, c, z)$ is called the hypergeometric function. The hypergeometric function takes a prominent position amongst the world of standard mathematical functions used in both pure and applied mathematics.

It should be noted that:

- when $a = 1$, $b = c$, or when the variables are all equal to 1, i.e. $a = b = c = 1$, (1.5) becomes the ordinary geometric series;
- setting $b = c = 1$ gives Newton's binomial series for $(1-z)^{-a}$;
- taking $a = b = 1/2$, $c = 3/2$ gives $\arcsin(\sqrt{z})/\sqrt{z}$, and
- when $a = b = 1$, $c = 2$, $z = 1$, it is the harmonic series

$$\begin{aligned} {}_2F_1(1, 1; 2; 1) &= \sum_{n=0}^{\infty} \frac{(1)_n(1)_n}{(2)_n}\frac{1^n}{n!} \\ &= \sum_{n=0}^{\infty} \frac{1}{n+1} = \sum_{n=1}^{\infty} = 1 + \frac{1}{2} + \frac{1}{3} + \dots . \end{aligned} \tag{1.6}$$

The name 'harmonic' derives from the concept of overtone, or harmonics in music: the wavelengths of the overtones of a vibrating string being 1/2, 1/3, 1/4 etc of the string's fundamental wavelength.

If the variable c, which occurs in the denominator of the series representation for the hypergeometric function, is zero or a negative integer, then the Gauss hypergeometric series is only defined if the zero is due to any one (or both) of the numerator variables occuring before the zero of the denominator variable. Otherwise—if the zero of the denominator variable precedes the zero of a numerator variable—the series is not defined.

On January 20, 1812, Carl Friedrich Gauss (1777–1855)[1] announced in his famous *Discquistiones Generales Cica Seriem Infinatum*, at the Paris International Congress of Mathematicians held in 1812, the hypergeometric functions not only as Euler's hypergeometric series, but also as the solutions of a second order ordinary differential equation throughout the complex plane. He discussed the convergence properties of the series and proved that it is convergent for $|z| < 1$, divergent for $|z| > 1$, and for $z = 1$ it is convergent if $\Re(c - a - b) > 0$ and divergent if

[1] More biographical details of Gauss, sometimes referred to as *Princeps mathematicorum* (Latin for 'the prince or foremost of mathematicians') and 'the greatest mathematician since antiquity' can be found in chapter 9; see also Srinivasa Rao and Van den Berghe (2004) and Tent (2008).

$\Re(c - a - b) < 0$. The approach of Gauss was new and he published only a part of his work in 1812. Gauss was aware of the multi-valuedness of the hypergeometric functions, known in recent times as the monodromy problem.

Gauss stated that the series, represented by the hypergeometric function (1.5), should be considered as a function of four variables a, b, c, z and *not* as a function of the variable z and parameters a, b, c. The modern notation for the Gauss hypergeometric function is

$$ {}_2F_1(a, b; c; z) \qquad \text{or} \qquad {}_2F_1\left(\begin{matrix} a, b \\ c \end{matrix}; z\right), \tag{1.7} $$

according to Barnes (1908). The latter notation is preferred for the generalized hypergeometric series:

$$ {}_qF_p\left(\begin{matrix} a_1, a_2, \ldots a_p \\ b_1, b_2, \ldots b_q \end{matrix}; z\right) = \sum_{k=0}^{\infty} \frac{(a_1)_k(a_2)_k \cdots (a_q)_k}{(b_1)_k(b_2)_k \cdots (b_p)_k} \frac{z^k}{k!}, \tag{1.8} $$

where p and q are integers in the applications to be presented later in this monograph, $q = p + 1$, with the extra numerator variable being a negative integer $-n$, is for terminating generalized hypergeometric series.

The hypergeometric series or Gauss series is thus:

$$ {}_2F_1(a, b; c; z) = \sum_{n=0}^{\infty} \frac{(a)_n(b)_n}{(c)_n} \frac{z^n}{n!} \tag{1.9} $$

where the prefix 2 and the suffix 1 of the ${}_2F_1(a, b; c; z)$ refer to the two variables in the numerator (a, b) and to the single denominator variable c.

It has been noted (Srinivasa Rao 1981) that the hypergeometric series has the folded-form of Horner for the polynomial evaluation built-in:

$$ 1 + \frac{ab}{c}\frac{z}{1}\left(1 + \frac{(a+1)(b+1)}{(c+1)}\frac{z}{2}\left(1 + \frac{(a+2)(b+2)}{(c+2)}\frac{z}{3}(1 + \ldots)\right)\right) \tag{1.10} $$

and it is useful for numerical computation, when any one of the numerator variables is a negative integer $-n$, and the hypergeometric series becomes a polynomial of degree n. For example, when $b = -2$, (1.5) reduces to:

$$ 1 + \frac{ab}{c}\frac{z}{1!} + \frac{a(a+1)b(b+1)}{c(c+1)}\frac{z^2}{2!} = 1 - \frac{2a}{c}\frac{z}{1!} + \frac{2a(a+1)}{c(c+1)}\frac{z^2}{2!} \tag{1.11} $$

a polynomial of degree $n = 2$, since all the other terms are zero. That is: $(-n)_{n+1} = (-2)_3 = 0$.

Therefore, whenever a numerator variable is a negative integer, $-n$, the series becomes a polynomial of the degree n and the folded-form is Horner's well-known method polynomial evaluation (Horner 1819). The finite geometric series (1.5), with n terms, in the Horner method can be written as:

$$ (1 + x(1 + x(1 + \cdots + x(1 + x)))), \tag{1.12} $$

and should be numerically evaluated inside out. The hypergeometric series has this Horner scheme built-in and, therefore, can be written as:

$$
\begin{aligned}
&{}_2F_1(a, b; c; z) \\
&= 1 + \frac{ab}{c}\frac{z}{1!}\left(1 + \frac{(a+1)(b+1)}{(c+1)}\frac{z}{2}\left(1 + \frac{(a+2)(b+2)}{(c+2)}\frac{z}{3} + \dots\right)\right).
\end{aligned} \tag{1.13}
$$

The number of additions in both (1.5) and (1.10) is n, but the number of multiplications reduces from $(n-1)n/2$, or $O(n^2)$ for large $n \to \infty$, resulting in a reduction in the computational time required for the numerical evaluation of the generalized hypergeometric series (see chapter 6).

For the terminating hypergeometric series, Euler (1748) obtained the famous transformation:

$$ {}_2F_1(a, -n; c; z) = (1-z)^{c+n-a}\,{}_2F_1(c-a, c+n; c; z). \tag{1.14} $$

For $z = 1$, Vandermonde[2] discovered the theorem in 1770, as an extension of the binomial theorem:

$$ {}_2F_1(a, -n; c; 1) = \frac{(c-a)(c-a+1)(c-a+2)\cdots(c-a+n-1)}{c(c+1)(c+2)\cdots(c+n-1)}, $$

or,

$$ {}_2F_1(a, -n; c; 1) = \frac{(c-a)_n}{(c)_n}, \tag{1.15} $$

which can be considered as an extension of the binomial theorem, where

$$ (\alpha)_n = \alpha(\alpha+1)(\alpha+2)\cdots(\alpha+n-1) = \frac{\Gamma(\alpha+n)}{\Gamma(\alpha)_n}, \qquad (\alpha)_0 = 1, \tag{1.16} $$

the Pochhammer symbol (1.4) is given in terms of the Gamma function, using the property of the Gamma function:

$$ \Gamma(n+1) = n\Gamma(n) = n!, \tag{1.17} $$

also called a rising factorial.

Gauss proved the famous summation theorem:

$$ {}_2F_1(a, b; c; 1) = \frac{\Gamma(c)\Gamma(c-a-b)}{\Gamma(c-a)\Gamma(c-b)}, \tag{1.18} $$

which yields the summation theorem of Vandermonde, for $b = -n$:

$$ {}_2F_1(a, -n; c; 1) = \frac{\Gamma(c)\Gamma(c-a+n)}{\Gamma(c-a)\Gamma(c+n)} = \frac{(c-a)_n}{(c)_n}. \tag{1.15} $$

[2] This theorem is also called the Chu–Vandermonde identity. See Askey (1975), pages 59–60.

The Gauss summation theorem is a direct consequence of an integral representation for the hypergeometric function, due to Euler (1748):

$$ {}_2F_1(a, b; c; z) = \frac{\Gamma(c)}{\Gamma(b)\Gamma(c-b)} \int_0^{\infty} t^{b-1}(1-t)^{a-c}(1-tz)^{-a}\,dt, \tag{1.19}$$
$$\Re(c) > \Re(b) > 0,$$

which, when $z = 1$, reduces to the beta integral:

$$\int_0^1 t^{b-1}(1-t)^{c-a-b-1}\,dt = B(b, c-a-b) = \frac{\Gamma(b)\Gamma(c-a-b)}{\Gamma(c-a)}. \tag{1.20}$$

Substituting (1.20) into (1.19) yields the Gauss summation theorem (1.18).

Gauss gave many relations between two or more of the Gauss hypergeometric series.

From the definition (1.9), it follows immediately that:

$$ {}_2F_1(a, b; c; z) = {}_2F_1(b, a; c; z). \tag{1.21}$$

1.2 The Gauss differential equation

After Wallis' discovery, for the next 200 years many other mathematicians studied similar series', most notably the Swiss mathematician Leonhard Euler, who gave, amongst many other results, the famous relation:

$$ {}_2F_1(-n, b; c; z) = (1-z)^{n+c-b}\,{}_2F_1(c+n, c-b; c; z). \tag{1.22}$$

From the Ratio test for the convergence of a Gauss series (1.9):

$$\lim_{n\to\infty}\left|\frac{u_{n+1}}{u_n}\right| = \left|\lim_{n\to\infty}\frac{\left(1+\frac{a}{n}\right)\left(1+\frac{b}{n}\right)}{\left(1+\frac{c}{n}\right)\left(1+\frac{1}{n}\right)}z\right| = |z|, \tag{1.23}$$

assuming that $a, b, c, \neq 0$ or negative integers, it follows that the series:

(i) converges for all $|z| < 1$
(ii) diverges for all $|z| > 1$ and
(iii) for $|z| = 1$, converges if $\Re(c-a-b) > 0$.

As mentioned earlier, Gauss regarded the ${}_2F_1(a, b; c; z)$ as a function of four variables, rather than as a series in z, and in a note added on February 10, 1812, Gauss (Gauss 1876) gave a remarkably full discussion on the convergence properties of such a series.

In the opinion of Slater (1966): 'during the next 40 years, the Göttingen School under C F Hindenberg wasted much effort on various complicated extensions of the binomial and multinomial theorems.'

Gauss established 15 relations between six contiguous functions:

$$ {}_2F_1(a \pm 1, b; c; z) \equiv F(a \pm 1), \quad F(b \pm 1), \quad F(c \pm 1) \tag{1.24}$$

and showed that there exists between any two of these contiguous functions linear relations with coefficients, at the most linear in z.

Gauss also proved a theorem now referred to as the (Gauss) summation theorem (1812):

$$ {}_2F_1(a, b; c; 1) = \frac{\Gamma(c)\Gamma(c-a-b)}{\Gamma(c-a)\Gamma(c-b)}, \quad \Re(c-a-b) > 0 \tag{1.25} $$

and gave many relations between two or more of these series. This result is easily proved from the Euler integral representation (1.19) for the Gauss function, which, for $z = 1$, becomes:

$$ {}_2F_1(a, b; c; 1) = \frac{\Gamma(c)}{\Gamma(b)\Gamma(c-b)} \int_0^1 t^{b-1}(1-t)^{c-a-b-1}\, dt \tag{1.26} $$

where the integral can be recognized as the beta function, which has the property:

$$ B(m, n) = \int_0^1 t^{m-1}(1-t)^{n-1}\, dt = \frac{\Gamma(m)\Gamma(n)}{\Gamma(m+n)}, \quad \Re(m) > 0, \Re(n) > 0, \tag{1.27} $$

so that

$$ {}_2F_1(a, b; c; 1) = \frac{B(b, c-a-b)}{B(b, c-b)} = \frac{\Gamma(c)\Gamma(c-a-b)}{\Gamma(c-a)\Gamma(c-b)} \tag{1.28} $$

for $c - a - b > 0$ and for all a, b, c complex (even without $\Re(c - a - b) > 0$).

The hypergeometric function is best defined as the solution of the second order ordinary differential equation (ODE):

$$ z\,(1-z)\,\frac{d^2y}{dz^2} + [c - (a+b+1)\,z]\,\frac{dy}{dz} - a\,b\,y = 0. \tag{1.29} $$

It was Kummer (1975; 1810–98) who showed that the Gauss ODE, characterized by three regular singular points at 0, 1, ∞, has one solution, which is ${}_2F_1(a, b; c; z)$. It is one of 24 solutions (see appendix A) all in terms of Gauss functions with variables, which are linear combinations of a, b, c, z. Kummer published a set of six distinct solutions of the hypergeometric equation. Each of these six solutions has four different forms, related to one another by Euler's transformations:

$$ F(a, b; c; z) = (1-z)^{-a}\, F\left(a, c-b; c; \frac{z}{z-1}\right) \tag{1.30} $$

$$ F(a, b; c; z) = (1-z)^{-b}\, F\left(c-a, b; c; \frac{z}{z-1}\right) \tag{1.31} $$

$$ F(a, b; c; z) = (1-z)^{c-a-b}\, F(c-a, c-b; c; z) \tag{1.32} $$

giving 24 forms in all. Often these forms are referred to as the Kummer solutions of the Gauss hypergeometric ODE. Another way of grouping them follows the four solutions given in each of the six regions of the variable:

$$z, \frac{1}{z}, 1 - z, \frac{1}{1 - z}, \frac{z}{1 - z}, \frac{1 - z}{z}, \tag{1.33}$$

with $0 < z < 1$. Together the 24 solutions cover the entire complex plane, or, 48 solutions if one includes the trivial mirror symmetry:

$$F(a, b; c; z) = F(b, a; c; z). \tag{1.34}$$

Though the factors in the differential equation may look rather strange, they are perfectly adapted to the use of their solutions in a wide variety of situations. General homogeneous linear second order differential equations are formally written as:

$$y'' + P(z)\, y' + Q(z)\, y = 0, \tag{1.35}$$

where y' and y'' are the first and second derivatives of $y$with respect to z and $P(z)$ and $Q(z)$ are:

$$P(z) = \frac{c - (a + b + 1)z}{z(1 - z)} \quad \text{and} \quad Q(z) = -\frac{ab}{z(1 - z)} \tag{1.36}$$

so that $z = 0$ and $z = 1$ are the only sigular points.

$$\begin{aligned} zP(z) &= \frac{c - (a + b + 1)z}{(1 - z)} = [c - (a + b + 1)z]\,(1 + z + z^2 + \cdots) \\ &= c + [c - (a + b + 1)\, z] + \cdots \end{aligned} \tag{1.37}$$

and

$$z^2\, Q(z) = -\frac{a\, b\, z}{(1 - z)} = -a\, b\, z\,(1 + z + z^2 + \cdots) = -a\, b\, z - a\, b\, z^2 - \cdots. \tag{1.38}$$

Thus, $z = 0$ is a *regular* singular point.

Definition: a singular point z_0 of (1.29) is said to be *regular* if the functions $(z - z_0)\, P(z)$ and $(z - z_0)^2\, Q(z)$ are analytic and *irregular* otherwise.

Similarly, since

$$(1 - z)\, P(z) = \frac{c - (a + b + 1)z}{z} = -(a + b + 1) + \frac{c}{z} \tag{1.39}$$

and

$$(1 - z)^2\, Q(z) = -(1 - z)\frac{ab}{z} = ab - \frac{ab}{z}, \tag{1.40}$$

$z = 1$ is also a regular singular point. The hypothesis is that $z\,P(z)$ and $z^2\,Q(z)$ are analytic at $z = 0$ and therefore have the power series expansions:

$$z\,P(z) = \sum_{n=0}^{\infty} p_n\, z^n \quad \text{and} \quad z^2\,Q(z) = q_n\, z^n, \tag{1.41}$$

which are valid on an interval $|z| < R$ for some $R > 0$.

Differentiating the solution in the series:

$$y' = \sum_{n=0}^{\infty} a_n(m+n)z^{m+n-1} \tag{1.42}$$

$$\begin{aligned} y'' &= \sum_{n=0}^{\infty} a_n(m+n)(m+n-1)z^{m+n-2} \\ &= z^{m-2}\sum_{n=0}^{\infty} a_n(m+n)(m+n-1)z^n. \end{aligned} \tag{1.42a}$$

The terms $P(z)\,y'$ and $Q(z)\,y$ can be written as:

$$\begin{aligned} P(z)\,y' &= \frac{1}{z}\left(\sum_{n=0}^{\infty} p_n z^z\right)\left(\sum_{n=0}^{\infty} a_n\,(m+n)\,z^{m+n-1}\right) = z^{m-2}\left(\sum_{n=0}^{\infty} a_n(m+n)x^n\right) \\ &= z^{m-2}\sum_{n=0}^{\infty}\left[\sum_{k=0}^{n} p_{n-k}a_k\,(m+k)\right]z^m \\ &= z^{m-2}\sum_{n=0}^{\infty}\left[\sum_{k=0}^{n-1} p_{n-k}a_k(m+k) + p_0\,a_n(m+n)\right]z^n \end{aligned} \tag{1.42b}$$

$$\begin{aligned} Q(z)\,y &= \frac{1}{z^2}\left(\sum_{n=0}^{\infty} q_n\,z^n\right)\left(\sum_{n=0}^{\infty} a_n z^{m+n}\right) = z^{m-2}\left(\sum_{n=0}^{\infty} q_n x^n\right)\left(\sum_{n=0}^{\infty} a_n z^n\right) \\ &= z^{m-2}\sum_{n=0}^{\infty}\left[\sum_{k=0}^{n} q_{n-k}\,a_k\right]z^k = z^{m-2}\sum_{n=0}^{\infty}\left[\sum_{k=0}^{n-1} q_{n-k}a_k + q_0\,a_n\right]z^k. \end{aligned} \tag{1.42c}$$

When these expressions (a), (b), (c) for y'', $P(z)\,y'$, $Q(z)\,y$ are substituted into the differential equation (1.29), it results in:

$$\begin{aligned} \sum_{n=0}^{\infty}\Bigg\{ & a_n\,[(m+n)(m+n-1) + (m+n)\,p_0 + q_0] \\ & + \sum_{k=0}^{n-1} a_k\,[(m+k)p_{n-k} + q_{n-k}]\Bigg\}\,z^n = 0 \end{aligned} \tag{1.43}$$

and equating to zero the coefficient of z^n yields the recursion formula for a_n:

$$a_n[(m+n)(m+n-1)+(m+n)\,p_0+q_0] + \sum_{k=0}^{n-1} a_k\,[(m+k)p_{n-k}+q_{n-k}] = 0, \tag{1.44}$$

which for $n = 0$ yields:

$$a_0\,[m(m-1)] + mp_0 + q_0] = a_0\,f(m) = 0. \tag{1.45}$$

Since $a_0 \neq 0, f(m) = 0$, or, equivalently:

$$m\,(m-1) + m\,p_0 + q_0 = 0, \tag{1.46}$$

which is the *indicial equation*. Its roots are m_1 and m_2, which are the possible values for m in the assumed solution. Also, they are the *exponents* of the differential equation at the regular singular point $z = 0$. For the Gauss second order ODE, since $p_0 = c$ and $q_0 = 0$, the indicial equation becomes:

$$m(m-1) + m\,c = 0, \qquad \text{or,} \qquad m\,(m-1+c) = 0. \tag{1.47}$$

Therefore, the exponents are:

$$m_1 = 0 \quad \text{and} \quad m_2 = 1 - c. \tag{1.48}$$

Assuming a power series solution (the Frobenius method) to the equation:

$$y = z^\lambda \sum_{n=0}^{\infty} a_n\,z^n = \sum_{n=0}^{\infty} a_n\,z^{\lambda+n}, \tag{1.49}$$

the exponent λ and coefficients a_n are to be determined.

Differentiating:

$$y' = \sum_n a_n(\lambda+n)z^{\lambda+n-1} = z^{\lambda-1}\sum_n a_n\,(\lambda+n)\,z^n \tag{1.50}$$

$$y'' = \sum_n a_n\,(\lambda+n)\,(\lambda+n-1)z^{\lambda+n-2} = z^{\lambda-2}\sum_n a_n(\lambda+n)\,(\lambda+n-1)\,z^n. \tag{1.51}$$

Substituting these into the Gauss ODE, we get:

$$z^{\lambda-2}z(1-z)\sum_n a_n(\lambda+n)(\lambda+n-1)\,z^n + [c-(a+b+1)\,z]z^{\lambda-1}\sum_n a_n\,(\lambda+n)\,z^n - a\,b\,z^\lambda \sum_n a_n z^n = 0. \tag{1.52}$$

Since $|z| < 1$, the coefficients of z must vanish term-by-term. The coefficient of the lowest power $z^{\lambda-1}$ determines λ and the general term $z^{\lambda+n}$, equating to zero yields a recursion formula for a_n. The coefficient of $z^{\lambda-1}$ gives rise to the indicial equation:

$$a_0\,\lambda\,(\lambda - 1) + c\,\lambda\,a_0 = [\lambda(\lambda - 1) + c\,\lambda]\,a_0 = 0. \tag{1.53}$$

Since $a_0 \neq 0$,

$$\lambda\,(\lambda - 1 + c) = 0 \quad \Rightarrow \quad \lambda = 0, \quad \text{or}, \quad \lambda = 1 - c \tag{1.54}$$

and the coefficient of $z^{\lambda+n}$ is:

$$\begin{aligned} &a_{n+1}\,(\lambda + n + 1)\,(\lambda + n) - a_n\,(\lambda + n)\,(\lambda + n - 1) \\ &\quad + c\,a_{n+1}(\lambda + n + 1) - (a + b + 1)\,a_n\,(\lambda + n) - a\,b\,a_n = 0 \end{aligned}$$

i.e. $(\lambda + n + 1)(\lambda + n + c)a_{n+1} - (\lambda + n)(a + b + \lambda + n)\,a_n - a\,b\,a_n = 0.$

Therefore,

$$a_{n+1} = \frac{(\lambda + n)(\lambda + n + a + b) + ab}{(\lambda + n + 1)(\lambda + n + c)}a_n = \frac{(\lambda + n + a)(\lambda + n + b)}{(\lambda + n + 1)(\lambda + n + c)}a_n. \tag{1.55}$$

Substituting $\lambda = 0$:

$$a_{n+1} = \frac{(a + n)(b + n)}{(c + n)(1 + n)}a_n \tag{1.56a}$$

or

$$a_n = \frac{(a + n - 1)(b + n - 1)}{(c + n - 1)n}a_{n-1}. \tag{1.56b}$$

Continuing this recursion procedure:

$$\begin{aligned} a_n &= \frac{(a + n - 1)(b + n - 1)}{(c + n - 1)n}\frac{(a + n - 2)(b + n - 2)}{(c + n - 2)(n - 1)}a_{n-2} \\ &= \frac{(a + n - 1)(b + n - 1)}{(c + n - 1)n}\frac{(a + n - 2)(b + n - 2)}{(c + n - 2)(n - 1)}a_{n-2}\cdots \\ &\quad \times \frac{(a + 1)(b + 1)}{(c + 1)(1 + 1)}\frac{a \cdot b}{c \cdot 1}a_0 = \frac{(a)_n(b)_n}{(c)_n(1)_n}z^n. \end{aligned} \tag{1.57}$$

Setting $a_0 = 1$, we obtain the Gauss series:

$$y_1 = 1 + \frac{ab}{c}z + \frac{a(a + 1)b(b + 1)}{c(c + 1)}\frac{z^2}{2!} + \cdots = \sum_{n=0}^{\infty}\frac{(a)_n\,(b)_n}{(c)_n\,(1)_n}z^n \tag{1.58}$$

and, the series solution in contemporary function notation due to Barnes (1908):

$$y_1 = {}_2F_1(a, b; c; z) \equiv F(a, b, c, z) \text{ of Gauss.}$$

The second independent solution for $\lambda = 1 - c$ results in the recurrence relation:

$$\begin{aligned} a_{n+1} &= \frac{(1 - c + n)(a + b + 1 - c + n) + ab}{(1 + n)(2 - c + n)} a_n \\ &= \frac{(a - c + n + 1)(b - c + 1 + n)}{(1 + n)(2 - c + n)} a_n, \end{aligned} \tag{1.59}$$

which yields the solution:

$$y_2 = z^{1-c} {}_2F_1(a - c + 1, b - c + 1; 2 - c; z). \tag{1.60}$$

Combining these two independent solutions, the general solution valid near the origin $z = 0$ is:

$$y = A\ {}_2F_1(a, b; c; z) + Bz^{1-c} {}_2F_1(a - c + 1, b - c + 1; 2 - c; z) \tag{1.61}$$

provided that $1 - c$ is not zero or a positive integer.

If $c = 1$, the two solutions become identical to ${}_2F_1(a, b; 1; z)$. Writing

$$y_1(z) = {}_2F_1(a, b; c; z) \tag{1.62}$$

and setting

$$y_2(z) = y_1(z) \log z + \sum_{n=1}^{\infty} a_n z^n \tag{1.63}$$

where the coefficients a_n can be determined by direct substitution of this solution to the differential equation (with $c = 1$).

A similar procedure holds when $1 - c$ is a positive integer.

Case 1: $z = 1$. Let $\xi = 1 - z$. The Gauss hypergeometric second order ODE becomes:

$$\xi(1 - \xi)\frac{d^2y}{d\xi^2} + [a + b - c + 1 - (a + b + 1)\xi]\frac{dy}{d\xi} - ab\ y = 0, \tag{1.64}$$

which is identical to the original equation with c replaced with $a + b - c + 1$ and $\xi = 1 - z$. In this case, the indicial equation becomes:

$$\lambda(\lambda + a + b - c) = 0 \tag{1.65}$$

so that the exponents are:

$$\lambda = 0 \quad \text{and} \quad \lambda = c - a - b. \tag{1.66}$$

The Gauss equation can be written in the normal form as:

$$y'' + \left[\frac{c}{z(1 - z)} - \frac{a + b + 1}{1 - z}\right]y' - \frac{a\ b}{z(1 - z)}y = 0 \tag{1.67}$$

or, equivalently as

$$z(1 - z)\, y'' + [c - (a + b + 1)z]\, y' - a\, b\, y = 0, \tag{1.68}$$

from which $z = 0, 1$ are seen to be the regular singular points of the equation. If z is replaced by $1/z$ then ∞ is also a regular singularity of the Gauss equation. It was Kummer (1836) who first showed how the Gauss second order ODE is satisfied by (1.9), which we repeat here

$${}_2F_1(a, b; c; z) = \sum_{n=0}^{\infty} \frac{(a)_n (b)_n}{(c)_n} \frac{z^n}{n!}, \tag{1.9}$$

and that there exist 24 solutions valid in any region of the complex z-plane, provided that a, b, c are not integers or zero (Slater 1966, Bell 1968, for all 24 solutions). Gauss gave four solutions in his 1812 paper, and Kummer obtained the four solutions of Gauss along with four each in five other regions of the complex z plane. The six regions corresponding to the variable:

$$z, \quad \frac{1}{z}, \quad 1 - z, \frac{1}{1 - z}, \frac{z}{1 - z}, \frac{z - 1}{z} \tag{1.69}$$

cover the entire complex plane, which is reiterated.

Case 2: $z = \infty$. For the point at ∞, the solution can be determined using the Laurent series, with $\xi = 1/z$:

$$y = \sum_{n=0}^{\infty} a_n\, z^{-\lambda - n}, \tag{1.70}$$

which makes the Gauss equation:

$$\begin{aligned}(1 - z)\sum_{n=0}^{\infty} a_n (n + \lambda)\, (n + \lambda + 1) z^{-n-\lambda-1} - [c - (a + b + 1)\, z] \\ \times \sum_{n=0}^{\infty} (n + \lambda) a_n z^{-n-\lambda-1} - ab \sum_{n=0}^{\infty} a_n z^{-n-\lambda} = 0.\end{aligned} \tag{1.71}$$

Equating the coefficient of $z^{-\lambda}$ to zero, yields the indicial equation:

$$\begin{aligned}&- a_0\, \lambda(\lambda + 1) + (a + b + 1)\, \lambda\, a_0 - ab\, a_0 = 0 \\ &\Rightarrow\; - [\lambda(\lambda + 1) - (a + b + 1)\, \lambda + ab]\, a_0 = 0.\end{aligned} \tag{1.72}$$

Since $a_0 \neq 0$, it follows that:

$$\begin{aligned}&\lambda(\lambda + 1) - (a + b + 1)\, \lambda + ab = (\lambda - a - b) + ab = 0 \\ \text{i.e.}\quad &\lambda^2 - \lambda(a + b) + ab = (\lambda - a)(\lambda - b) = 0.\end{aligned} \tag{1.73}$$

So, the exponents are:

$$\lambda = a, \qquad \text{and} \qquad \lambda = b. \tag{1.74}$$

Equating the coefficients of $z^{-n-\lambda-1}$ we get:

$$\begin{aligned} &a_n(n+\lambda)(n+\lambda+1) - a_{n+1}(n+\lambda+1)(n+\lambda+2) \\ &\quad - c(n+\lambda)a_n + (a+b+1)(n+\lambda+1)a_{n+1} + a\,b\,a_{n+1} = 0 \\ \text{i.e.}\quad &a_n(n+\lambda)(n+\lambda-c+1) \\ &\quad = a_{n+1}[(n+\lambda+1)(n+\lambda+2-a-b-1) + a\,b], \end{aligned} \tag{1.75}$$

which results in the recurrence relation:

$$a_{n+1} = \frac{(n+\lambda)(n+\lambda-c-1)}{(n+\lambda+1)(n+\lambda-a-b+1)+ab}a_n$$

or

$$a_{n+1} = \frac{(n+\lambda)(n+\lambda-c+1)}{(n+\lambda-a+1)(n+\lambda-b+1)}a_n. \tag{1.76}$$

Corresponding to the root $\lambda = a$:

$$a_{n+1} = \frac{(n+a)(n+a-c+1)}{(n+1)(n+a-b+1)}a_n. \tag{1.77}$$

Taking $a_0 = 1$, the solution becomes:

$$y_1 = z^{-a}{}_2F_1\left(a, a-c+1; a-b+1; \frac{1}{z}\right). \tag{1.78}$$

Corresponding to the root $\lambda = b$ we have:

$$a_{n+1} = \frac{(n+b)(n+b-c+1)}{(n+1)(n-a+b+1)}a_n \tag{1.79}$$

so that

$$y_2 = z^{-b}{}_2F_1\left(a, a-c+1; a-b+1; \frac{1}{z}\right). \tag{1.80}$$

Therefore, the required general solution is a linear combination of the two independent solutions:

$$\begin{aligned} y = {} &Az^{-a}{}_2F_1\left(a, a-c+1; a-b+1; \frac{1}{z}\right) \\ &+ Bz^{-b}{}_2F_1\left(b, b-c+1; b-a+1; \frac{1}{z}\right). \end{aligned} \tag{1.81}$$

Summarizing, the regular singular points of the Gauss hypergeometric equation are:

(i) $z = 0$ with exponents $0, 1-c$;
(ii) $z = \infty$ with exponents a, b;
(iii) $z = 1$ with exponents $0, c-a-b$.

These facts are exhibited symbolically by denoting the most general solution of the Gauss second order ODE by a scheme of the form:

$$y = P\left\{\begin{matrix} 0 & \infty & 1 & \\ 0 & a & 0 & z \\ 1-c & b & c-a-b & \end{matrix}\right\}, \tag{1.82}$$

which is called the *Riemann-P-function*[3] of the hypergeometric equation (1857). This is also called the Papperitz equation (1885).

The crucial facts to note are the following general features of the hypergeometric equation: the coefficients of y'', y', y in the ODE:

$$z\,(1-z)\,\frac{d^2y}{dz^2} + [c - (a+b+1)\,z]\,\frac{dy}{dz} - a\,b\,y = 0 \tag{1.83}$$

or

$$(\theta\,(\theta + c - 1) - z\,(\theta + a)\,(\theta + b))\,y = 0 \tag{1.84}$$

where $\theta \equiv z\,(d/dz)$, are polynomials of degree 2, 1 and 0, and also that the first of these polynomials has distinct real zeros ($z = 0$) and ($z = 1$). Any differential equation with these characteristics can be brought into the hypergeometric form by a linear change of the independent variable, and hence can be solved near its singular points in terms of the hypergeometric function. To make the point clear, consider the generalized differential equation of the type:

$$(x - A)\,(x - B)\,y'' + (C + D\,x)\,y' + E\,y = 0, \qquad A \neq B. \tag{1.85}$$

If we change the independent variable from x to t by:

$$t = \frac{x - A}{B - A}, \tag{1.86}$$

then $x = A$ corresponds to $t = 0$ and $x = B$ corresponds to $t = 1$.

The hypergeometric equation now takes the form:

$$t(1-t)\,y_t'' + (F + G\,t)\,y_t' + H\,y = 0 \tag{1.87}$$

where $(C + A)/(B - \mathrm{A})$, $G = D$, $H = E$. This is a hypergeometric equation with a, b, c defined through $F = c$, $G = -(a + b + 1)$, $H = -\,a\,b$, and can therefore be solved near $t = 0$ and $t = 1$ in terms of the hypergeometric function; which means the original equation can be solved in terms of the same function near $x = A$ and $x = B$.

[3] Named after George Friedrich Bernhard Riemann 1826–66, German mathematician. For a fascinating biography see Derbyshire (2003). See also: www-groups.dcs.st-and.ac.uk/history/Biographies/Riemann.html.

1.3 Special functions

In this monograph after introducing the hypergeometric series and providing the basics of group theory in the first two chapters, we will present applications of group theory to the study of:

- the 24 Kummer solutions of the Gauss differential equation;
- the discovery of a new 72 element group for the ${}_3F_2(a, b, c; d, e; 1)$ transformation of Weber–Erdelyi;
- the group theory of the terminating and non-terminating ${}_3F_2(1)$ transformations;
- angular momentum and the rotation group;
- angular momentum recoupling and sets of ${}_4F_3(1)$s;
- double and triple hypergeometric series;
- beta integral method and hypergeometric transformations;
- Gauss, hypergeometric series and Ramanujan.

What is so special about the special functions? These functions are in general solutions to certain second order ordinary linear differential equations which occur in a number of problems in the mathematical sciences. Nicco Tamme says: (see page 9 of Lakshminarayanan and Varadharajan 2015)

> We call a special function 'special' when just as the logarithm exponential and trigonometric (the elementary functions) belongs to the tool box of the applied mathematician, the physicist and the engineer.

Richard Askey defined a special function as follows (in the preface of Askey *et al* 1984):

> Certain functions appear so often that it is convenient enough to give them names. These functions are collectively called special functions. There are many examples and no single way of looking at them can illustrate all examples or even all important properties of a single example of a special function.

The three volume 'Higher Transcendental Functions' also called the Batemann manuscript project (see Erdély *et al* 1953) has a list of over a 1000 special functions[4]. Many of these are nothing but special forms of hypergeometric series. The special functions named after Bessel, Hermite, Jacobi, Laguere, Legendre, Chebyshev are the ones frequently encountered in mathematical sciences. A more recent tabulation is the NIST handbook of Mathematical Functions by Olver, Lozier, Boisvert and Clark (2010). The reader is referred to the discussion in section 1.6 of Lakshminarayanan and Varadarajan (2015).

[4] The complete three volumes (updated) of the Batemann manuscript project is available online at: http://caltech.edu/43491.

Another classical ordinary differential equation of considerable importance is the *confluent hypergeometric equation*. It is obtained from the Gauss second order ODE:

$$z\,(1-z)\,y'' + [c-(a+b+1)\,z]\,y' - a\,b\,y = 0 \tag{1.88}$$

by changing the variable from z to $x = b\,z$. Then we have:

$$\begin{aligned} y' &= \frac{dy}{dz} = \frac{dy}{dx}\frac{dx}{dz} = b\frac{dy}{dx}; \\ y'' &= \frac{d}{dz}\frac{dy}{dz} = \left(b\frac{d}{dx}\,b\,\frac{dy}{dx}\right) = b^2\frac{d^2y}{dx^2} \end{aligned} \tag{1.89}$$

so that the equation becomes:

$$\frac{x}{b}\left(1-\frac{x}{b}\right)b^2\,\frac{d^2y}{dx^2} + \left[c-(a+b+1)\frac{x}{b}\right]b\,\frac{dy}{dx} - a\,b\,y = 0 \tag{1.90}$$

$$\text{i.e.}\quad x\left(1-\frac{x}{b}\right)\frac{d^2y}{dx^2} + \left[(c-x)-(a+1)\,\frac{x}{b}\right]\frac{dy}{dx} - a\,y = 0, \tag{1.91}$$

which has the singular points at $x = 0$; $x = b$; and $x = \infty$.

It differs from the Gauss second order ODE in that the singular point at $x = b$ is now mobile. If we let $b \to \infty$, then the transformed differential equation becomes:

$$x\frac{d^2y}{dx^2} + (c-x)\,\frac{dy}{dx} - a\,y = 0, \tag{1.92}$$

where $x = b\,z$. This equation is called the *Kummer equation* and can be obtained from the Gauss equation by replacing z with x/b and letting $b \to \infty$ since

$$\lim_{b\to\infty}\frac{(b)_n}{b^n} \to 1. \tag{1.93}$$

The singular point at b has coalesced with the one at ∞, and the *confluence of the two singular points at* ∞ produces an *irregular* singular point there. The function defined by the series is:

$$y = {}_1F_1(a;\,c;\,z)$$

or

$${}_1F_1(a,\,c,\,z) \equiv \lim_{b\to\infty}{}_2F_1\left(a,\,b;\,c;\,\frac{z}{b}\right) \tag{1.94}$$

and is known as the *confluent hypergeometric function*. Kummer's equation is referred to as the confluent hypergeometric equation.

The intimate relationship that exists between the hypergeometric function and the special functions has been enumerated as a theorem, by Bell (1968):

Theorem:

(1)
$$P_n(x) = {}_2F_1\left(-n, n+1; 1; \frac{1-x}{2}\right)$$
⇔ Legendre polynomial.

(2)
$$P_n^m(x) = \frac{(n+m)!}{(n-m)!}\frac{(1-x^2)^{m/2}}{2^m\, m!}$$
$$\times\, {}_2F_1\left(m-n, m+n+1; m+1; \frac{1-x}{2}\right)$$
⇔ associated Legendre polynomial.

(3)
$$J_n(x) = \frac{e^{-ix}}{n!}\left(\frac{x}{2}\right)^n {}_1F_1\left(n+\frac{1}{2}; 2n+1; 2\,i\,x\right)$$
⇔ Bessel function of the 1st kind.

(4)
$$H_{2n}(x) = (-1)^n \frac{(2n)!}{n!} {}_1F_1\left(-n; \frac{1}{2}; x^2\right)$$
$$H_{2n+1}(x) = (-1)^n \frac{2(2n+1)!}{n!} z\; {}_1F_1\left(-n; \frac{3}{2}; x^2\right)$$
⇔ Hermite polynomials.

(5) $L_n(x) = {}_1F_1(-n; ;1; z)$
⇔ Laguerre polynomial.

$$L_n^k(x) = \frac{\Gamma(n+k+1)}{n!\,\Gamma(k+1)} \times {}_1F_1(-n; ;1; z)$$
⇔ associated Laguerre polynomial.

(6)
$$T_n(x) = {}_2F_1\left(-n, n; \frac{1}{2}; \frac{1-x}{2}\right),$$
$$U_n(x) = \sqrt{1-x^2}\,n\; {}_2F_1\left(-n+1, n+1; \frac{3}{2}; \frac{1-x}{2}\right)$$
⇔ Tchebychev/Tchebycheff/Tschebycheff polynomial.

(7) $$C_n^{\lambda}(x) = \frac{\Gamma(n + 2\lambda)}{n!\,\Gamma(2\lambda)}\,{}_2F_1\left(-n,\, n + 2\lambda;\, \lambda + \frac{1}{2};\, \frac{1 - x}{2}\right)$$

$\Leftrightarrow$ Gegenbauer polynomial.

(8) $$P_n^{\alpha,\,\beta}(x) = \frac{\Gamma(n + \alpha + 1)}{n!\Gamma(\alpha)}\,{}_2F_1\left(-n,\, n + \alpha + \beta + 1;\, \alpha + 1;\, \frac{1}{2}(x - 1)\right)$$

$\Leftrightarrow$ Jacobi polynomial.

In each case, the result may be proved by expanding the hypergeometric function as a series, compared with a known series for the given function. Table 12.2 in Lakshminarayanan and Varadarajan (2015) gives a list of various special functions in terms of the hypergeometric functions, including many not described here (LV, chapter 12, pages 269–89). Also, their figure 21.1 gives a block diagram that summarizes succinctly the intimate relationship between the special functions and their corresponding hypergeometric functions. We illustrate this by proving the result (1) for the Legendre polynomial.

Proof of (1):

$$P_n(x) = {}_2F_1\left(-n,\, n + 1;\, 1;\, \frac{1 - x}{2}\right) \tag{1.95}$$

$\Leftrightarrow$ Legendre polynomial.

By definition

$${}_2F_1\left(-n,\, n + 1;\, 1;\, \frac{1 - x}{2}\right) = \sum_{r=0}^{\infty} \frac{(-n)_r(n + 1)_r}{(1)_r(1)_r}\left(\frac{1 - x}{2}\right)^r \tag{1.96}$$

where

$$\begin{aligned}(-n)_r &= (-n)(-n + 1)(-n + 2)\cdots(-n + r - 1)\\ &= (-1)^r n(n - 1)(n - 2)\cdots(n - r + 1)\\ &= (-1)^r\frac{n!}{(n - r)!}.\end{aligned} \tag{1.97}$$

Also,

$$(n + 1)_r = (n + 1)(n + 2)\cdots(n + r) = \frac{(n + r)!}{n!} \tag{1.98}$$

and $(1)_r = 1 \cdot 2 \cdots r = r!$ so that we have:

$$
\begin{aligned}
{}_2F_1\left(-n, n+1; 1; \frac{1-x}{2}\right) &= \sum_{r=0}^{n} \frac{n!}{(n-r)!} \frac{(n+r)!}{n!} \frac{1}{(r!)^2} \frac{(1-x)^r}{2^r} \\
&= \sum_{r=0}^{n} \frac{(-1)^r}{2^r} \frac{(n+r)!}{(n-r)!(r!)^2}(1-x)^r \qquad (1.99) \\
&= \sum_{r=0}^{n} \frac{(n+r)!}{2^r (n-r)!(r!)^2}(x-1)^r.
\end{aligned}
$$

To show that this is the same as $P_n(x)$, write $P_n(x)$ as a power series in $(x - 1)$. Using the Taylor series expansion:

$$P_n(x) = \sum_{r=0}^{n} P_n^{(r)} \frac{(x-1)^r}{r!} \qquad (1.100)$$

where $P_n^{(r)}(1)$ is the rth derivative of $P_n(x)$ evaluated at $x = 1$.

To calculate $P_n^{(r)}(1)$, use the generating function for the Legendre polynomials given by:

$$\frac{1}{(1 - 2tx + t^2)^{1/2}} = \sum_{n=0}^{\infty} P_n(x) t^n \qquad (1.101)$$

so that, by differentiating r times with respect to x, we get:

$$\sum_{n=0}^{\infty} P_n^{(r)}(x)\, t^n = \frac{d^r}{dx^r}(1 - 2tx + t^2)^{-1/2} \qquad (1.102)$$

$$
\begin{aligned}
&= (-2t)^r\left(-\frac{1}{2}\right)\left(-\frac{1}{2} - 1\right)\left(-\frac{1}{2} - 2\right)\cdots\left(-\frac{1}{2} - r + 1\right)(1 - 2tx +^2)^{-\frac{1}{2}-r} \\
&= 2^r\, t^r\, \frac{1}{2}\left(\frac{1}{2} + 1\right)\left(\frac{1}{2} + 2\right)\cdots\left(\frac{1}{2} + r - 1\right)(1 - 2tx + x^2)^{-\frac{1}{2}-r} \qquad (1.103) \\
&= t^r\, 1 \cdot 3 \cdot 5 \cdots (2r - 1)(1 - 2tx + t^2)^{-\frac{1}{2}-r} \\
&= t^r \frac{(2r)!}{2^r\, r!}(1 - 2tx + t^2)^{-\frac{1}{2}-r}.
\end{aligned}
$$

Setting $x = 1$, we get, by the binomial theorem:

$$\begin{aligned}
\sum_{n=0}^{\infty} P_n^{(r)}(1)t^n &= \frac{t^r\,(2r)!}{2^r\,r!}(1 - 2t + t^2)^{-\frac{1}{2}-r} = \frac{t^r\,(2r)!}{2^r\,r!}(1 - t)^{-1-2r} \\
&= \frac{t^r\,(2r)!}{2^r\,r!}\left\{1 + (1 + 2r)t + \frac{(1 + 2r)(2 + 2r)}{2!}t^2 \right. \\
&\quad \left. + (2r + 1)(2r + 2)(2r + 3)\frac{t^3}{3!} + \dots\right\} \\
&= \frac{t^r\,(2r)!}{2^r\,r!}\sum_{s=0}^{\infty}\frac{(2r + s)}{(2r)!\,s!} = \frac{1}{2^r\,r!}\sum_{s=0}^{\infty}\frac{(2r + s)!}{s!}t^{r+s}.
\end{aligned} \tag{1.104}$$

Replacing $r + s = n$, with $0 \leqslant s \leqslant \infty \Rightarrow r \leqslant n \leqslant \infty$:

$$\sum_{n=0}^{\infty} P_n^{(r)}(1)\; t^n = \frac{1}{2^r\,r!}\sum_{n=r}^{\infty}\frac{(r + n)!}{(n - r)!}t^n. \tag{1.105}$$

Equating the coefficient of t^n gives:

$$\begin{aligned}
P_n^{(r)}(1) &= \frac{1}{2^r\,r!}\frac{(r + n)!}{(n - r)!} \text{ for } n \geqslant r \\
&= 0 \quad \text{for } n < r.
\end{aligned} \tag{1.106}$$

Hence,

$$\begin{aligned}
P_n(r) &= \sum_{r=0}^{n} P_n^{(r)}(1)\frac{(x - 1)^r}{r!} = \sum_{r=0}^{\infty}\frac{1}{2^r\,r!}\frac{(n + r)!}{(n - r)!}\frac{(x - 1)^r}{r!} \\
&= \sum_{r=0}^{n}\frac{(n + r)!}{(n - r)!}\frac{(x - 1)^r}{(r!)^2} = {}_2F_1\!\left(-n,\, n + 1;\, 1;\, \frac{1 - x}{2}\right).
\end{aligned} \tag{1.107}$$

1.4 Properties of the hypergeometric functions

We present here some of the important properties of the Gauss hypergeometric function:

- It can be verified that:

$$\lim_{b\to\infty} {}_2F_1\!\left(a,\, b;\, a;\, \frac{z}{a}\right) = e^z. \tag{1.108}$$

Proof is based on the fact:

$$\begin{aligned}
\lim_{b\to\infty} = \frac{(b)_n}{b^n} &= \lim_{b\to\infty}\frac{b(b + 1)(b + 2)\cdots(b + n - 1)}{b^n} \\
&= \lim_{b\to\infty} 1\cdot\left(\frac{1}{b} + 1\right)\left(\frac{2}{b} + 1\right)\dots\left(\frac{n - 1}{b} + 1\right) = 1.
\end{aligned}$$

Therefore

$$\lim_{b\to\infty} {}_2F_1\left(a;\, b;\, a;\, \tfrac{z}{b}\right) = \lim_{b\to\infty} \sum_n \frac{(a)_n (b)_n}{(a)_n n!}\left(\frac{z}{b}\right)^n$$
$$= \sum_n \lim_{n\to\infty} \frac{(b)_n}{b_n}\frac{z^n}{n!} = \sum_n \frac{z^n}{n!} e^z.$$

Also,

$${}_1F_1(a;\, a;\, z) = {}_{-}F_{-}(-;-;z) = e^z. \tag{1.109}$$

- It can be verified that:

$$\sin(x) = x \lim_{b\to\infty} {}_2F_1\left(a,\, a;\, \frac{3}{2};\, -\frac{x^2}{4a^2}\right) = x \lim_{b\to\infty} \sum_n \frac{(a)_n (a)_n}{\left(\frac{3}{2}\right)_n}\left(-\frac{x^2}{4a^2}\right)^n$$
$$= x \lim_{b\to\infty} \sum_n \frac{(a)_n}{a_n}\frac{(a)_n}{a_n}\frac{(-1)^n}{\left(\frac{3}{2}\right)_n}\frac{x^{2n}}{4^n n!} = \sum_n \frac{(-1)^n\, x^{2n+1}}{\left(\frac{3}{2}\right)_n 2^{2n}\, n!}. \tag{1.110}$$

Since

$$\left(\frac{3}{2}\right)_n 2^{2n}\, n! = \frac{3}{2}\left(\frac{3}{2}+1\right)\left(\frac{3}{2}+2\right)\cdots\left(\frac{3}{2}+n-1\right)2^{2n}\, n!$$
$$= 3\cdot 5\cdot 7\cdots(3+(2n-2))\,\frac{2^{2n}}{2^n}\, n! \tag{1.111}$$
$$= 3\cdot 5\cdot 7\cdots(2n+1)\cdot(1\cdot 2\cdot 3\cdots n)\, 2^n$$
$$= 3\cdot 5\cdot 7\cdots(2n+1)\cdot 2\cdot 4\cdot 6\cdots(2n) = (2n+1)!.$$

Therefore,

$$\sin(x) = \sum_n (-1)^n \frac{x^{2n+1}}{(2n+1)!}, \tag{1.112}$$

which is the desired series expansion result for the trigonometric Sine function.

- It can be easily shown that:

$$\cos(x) = \lim_{a\to\infty} {}_2F_1\left(a,\, a;\, \frac{1}{2};\, -\frac{x^2}{4a^2}\right)$$
$$= \lim_{a\to\infty} \sum_n \frac{(a)_n (a)_n}{\left(\frac{1}{2}\right)_n n!}\left(-\frac{x^2}{4a^2}\right)^n$$
$$= \lim_{a\to\infty} \sum_n \frac{(a)_n}{a^n}\frac{(a)_n}{a^n}\frac{(-1)^n}{\left(\frac{1}{2}\right)_n}\frac{x^{2n}}{4^n\, n!} \tag{1.113}$$
$$= \sum_n \frac{(-1)^n}{\left(\frac{1}{2}\right)_n}\frac{x^{2n}}{2^{2n}\, n!}.$$

Since

$$\left(\frac{1}{2}\right)_n 2^{2n}\, n! = \frac{1}{2}\left(\frac{1}{2}+1\right)\left(\frac{1}{2}+2\right)\cdots\left(\frac{1}{2}+n-1\right)2^{2n}\, n!$$
$$= 1(1+2)(1+4)\cdots(1+(2n-2))\,\frac{2^{2n}}{2^n}\,n! \qquad (1.114)$$
$$= 1\cdot 3\cdot 5\cdots(2n-1)\cdot(1\cdot 2\cdot 3\cdots n)2^n$$
$$= 1\cdot 3\cdot 5\cdots(2n+1)\cdot 2\cdot 4\cdot 6\cdots(2n) = (2n)!.$$

Note. The general result is:

$$(a)_{2n} = \left(\frac{a}{2}\right)_n\left(\frac{a}{2}+\frac{1}{2}\right)_n 2^{2n}. \qquad (1.115)$$

Therefore,

$$\cos(x) = \sum_n (-1)^n \frac{x^{2n}}{(2n)!}, \qquad (1.116)$$

which is the desired series expansion result for the trigonometric Cosine function.

- It can be verified that:

$$_2F_1(1, b; b; x) = \sum_n \frac{(1)_n(b)_n}{(b)_n}\frac{x^n}{n!}$$
$$= 1 + x + x^2 + \cdots + x^n + \cdots = \frac{1}{1-x} = (1-x)^{-1}. \qquad (1.117)$$

- The general binomial theorem is obtained for:

$$_2F_1(-n, b; b; x) = \sum_r \frac{(-n)_r(b)_r}{(b)_r r!}x^r \qquad (1.118)$$

where

$$(-n)_r = (-n)(-n+1)(-n+2)\cdots(-n+r-1)$$
$$= (-1)^r\, n(n-1)(n-2)\cdots(n-r+1) \qquad (1.119)$$
$$= (-1)^r \frac{n!}{(n-r)!}.$$

Therefore, the binomial theorem, (1.15), is:

$$_2F_1(-n, b; b; x) = \sum_r (-1)^r \frac{n!}{r!\,(n-r)!}x^r = \sum_r {}^nC_r\,(-x)^r = (1-x)^n. \qquad (1.120)$$

- It is also easy to verify:

$$\begin{aligned}\log(1+x) &= x - \frac{x^2}{2} + \frac{x^3}{3} + \cdots = \sum_{n=0}^{\infty}(-1)^n \frac{x^{n+1}}{n+1} \\ &= x\ {}_2F_1(1,\ 1;\ 2;\ x) = x\sum_n \frac{(1)_n(1)_n}{(2)_n n!}x^n = \sum_n \frac{(1)_n}{(2)_n}x^n \\ &= \sum_n (-1)^n \frac{(1)_n}{(2)_n}x^{n+1} = \sum_n (-1)^n \frac{x^{n+1}}{n+1}\end{aligned} \quad (1.121)$$

since

$$\frac{(1)_n}{(2)_n} = \frac{1\cdot 2\cdot 3\cdots n}{2\cdot 3\cdot 4\ldots n(n+1)} = \frac{1}{n+1} \quad (1.122)$$

- For $x^2 < 1$, $-\pi/2 < \sin^{-1}x < /\pi 2$:

$$\begin{aligned}\sin^{-1}x &= x\ {}_2F_1\left(\frac{1}{2},\ \frac{1}{2};\ \frac{3}{2};\ x^2\right) \\ &= x + \frac{x^3}{2\cdot 3} + \frac{1\cdot 3\cdot x^5}{2\cdot 4\cdot 5} + \frac{1\cdot 3\cdot 5\cdot x^7}{2\cdot 4\cdot 6\cdot 7} + \cdots\end{aligned} \quad (1.123)$$

- For $x^2 < 1$:

$$\tan^{-1}x = x\ {}_2F_1\left(\frac{1}{2},\ 1;\ \frac{3}{2};\ -x^2\right) = x - \frac{x^3}{3} + \frac{x^5}{5} - \frac{x^7}{7} + \ldots. \quad (1.124)$$

We recall that while announcing the discovery of the second order ODE, characterized by three regular singular points at 0, 1 and ∞, Gauss made the significant observation that the hypergeometric function $F(a, b, c, z)$ should *not* be considered as a function of the variable z, and three parameters—viz a, b and c—but as a function of four variables, which in latter day notation, with credit to Barnes, became ${}_2F_1(a, b; c; z)$. This observation—along with the application of group theory for a study of the 24 Kummer (1836) solutions for the Gauss second order ordinary differential equation—is related to the discovery that the set of solutions has an elegant, simple group theoretic structure associated with the 24 symmetries of an ordinary cube (Lievens *et al* 2005).

Erdelyi (1939) showed that a recursive use of a ${}_3F_2(a, b, c; d, e; 1)$ transformation results in another transformation for ${}_3F_2(1)$. This recursive procedure is continued until it leads to a set of transformations that form a group (Srinivasa Rao *et al* 1992). These researchers then studied the group theory of the terminating and non-terminating transformations of ${}_3F_2(a, b, c; d, e; 1)$.

The Bailey transformation for a terminating ${}_4F_3(a, b, c, d; e, f, g; 1)$ can also be used recursively. Such a procedure resulted in a study of the group theory associated with the ${}_4F_3(1)$ transformations.

After presenting the basics of group theory—necessary for this monograph to be self-sufficient—we present the relationship of:

(i) the 3-j coupling coefficient in quantum theory of angular momentum and its relationship to the ${}_3F_2(1)$,
(ii) the recoupling 6-j coefficient of angular momentum and its relationship to the ${}_4F_3(1)$,
(iii) the 9-j or $\ell s - jj$ transformation coefficient and, at its simplest, knows the triple sum series form, its relationship to a triple hypergeometric series, and the realm of the study of special functions of mathematical physics.

To summarize, in the following chapters of this monograph, we will present the following.

- Group theory: basics.
- Group theory of the Kummer solutions of the Gauss differential equation.
- Group theory of terminating and non-terminating ${}_3F_2(a, b, c; d, e; 1)$ transformations.
- Angular momentum and the rotation group.
- Angular momentum recoupling and sets of ${}_4F_3(1)$.
- Symmetries of the 9-j coefficient and the triple hypergeometric series.
- The beta integral method and hypergeometric transformations.
- In the last chapter we highlight the work of Ramanujan on the hypergeometric series, who reconstructed (without a formal education beyond school level) all that was known in Europe at that time with only the hint of the Gauss summation theorem[5].

1.5 A conjecture

In this chapter, the relationships shown by Bell (1968) on special functions reveal the intimate relationship between the special functions and the hypergeometric functions. The Legendre function $P_n(x) = {}_2F_1(-n, n+1; 1; \frac{1-x}{2})$ has the four variables:

$$a = -n, \quad b = n + 1, \quad c = n, \quad z = \frac{1-x}{2}. \tag{1.125}$$

[5] Srinivasa Ramanujan (1887–1920), an Indian mathematician, had no formal training in pure math but made important contributions to number theory, infinite series, continued fractions, etc. Ramanujan's life has been celebrated in popular culture. For example, in the 2015 movie *The Man Who Knew Infinity* (based on a best selling biography Robert Kanigel 1991), a NOVA/PBS documentary *The Man Who Loved Numbers* [season 15, episode 19], plays, novels (Leavitt, 2008), etc. More details are given in chapter 9.

We change the variable in the Gauss second order ODE (1.67) from z to $\frac{1-x}{2}$ to get:

$$(1 - x^2)\, y'' - 2\, x\, y' + n\,(n + 1) = 0 \tag{1.126}$$

corresponding to the theorem (1) of Bell for $P_n(x)$.

In the confluent hypergeometric equation (1.88)

$$x\, y'' + (c - x)\, y' - a\, y = 0 \tag{1.127}$$

changing the variables $a = -n$, $c = 1$, results in:

$$x\, y'' + (1 - x)\, y' + n\, y = 0, \tag{1.128}$$

which is the theorem (5a) in Bell for the Laguerre polynomial.

Our conjecture is to find an answer to the question: is it possible to establish, *à la* Bell's theorem for special functions, a corresponding theorem for the differential equations satisfied by those functions?

On the optimistic note that this question will perhaps be answered soon, we conclude this chapter.

Appendix A

The Gauss second order ordinary differential equation, characterized by its three regular singular points at 0, 1, ∞ is:

$$z(1 - z)\,\frac{d^2y}{dz^2} + [(c - a - b)\, z - 1]\,\frac{dy}{dz} - a\, b\, y = 0.$$

Kummer (1836) discovered that it has 24 solutions and not just the mandatory two. They are:

$$y_1 = F(a, b; c; z)$$

$$y_2 = (1 - z)^{-b}\, F\left(b, c - a; c; \frac{z}{z - 1}\right)$$

$$y_4 = (1 - z)^{-a}\, F\left(c - b, a; c; \frac{z}{z - 1}\right)$$

$$y_5 = z^{1-c}\, F(1 + a - c, 1 + b - c; 1 + a + b - c; 1 - z)$$

$$y_6 = F(b, a; 1 + a + b - c; 1 - z)$$

$$y_7 = z^{-a}\, F\left(a, 1 + a - c; 1 + a + b - c; \frac{z - 1}{z}\right)$$

$$y_8 = z^{-b}\, F\left(1 + b - c, b; 1 + a + b - c; \frac{z - 1}{z}\right)$$

$$y_9 = z^{-a}\, F\left(1 + a - c, a; 1 + a - b; \frac{1}{z}\right)$$

$$y_{10} = z^{b-c}\,(1 - z)^{c-a-b}\, F\left(c - b, 1 - b; 1 + a - b; \frac{1}{z}\right)$$

$$y_{11} = (1 - z)^{-a} F\left(a, c - b; 1 + a - b; \frac{1}{1 - z}\right)$$
$$y_{12} = z^{1-c}(1 - z)^{c-a-1} F\left(1 - b, 1 + a - c; 1 + a - b; \frac{1}{1 - z}\right)$$
$$y_{13} = z^{-b} F\left(b, 1 + b - c; 1 + b - a; \frac{1}{z}\right)$$
$$y_{14} = z^{a-c}(1 - z)^{c-a-b} F(1 - a, c - a; 1 + c - a - b; 1 - z)$$
$$y_{15} = (1 - z)^{-b} F\left(c - a, b; 1 + b - a; \frac{1}{1 - z}\right)$$
$$y_{16} = z^{1-c}(1 - z)^{c-b-1} F\left(1 + b - c, 1 - a; 1 + b - a; \frac{1}{1 - z}\right)$$
$$y_{17} = (1 - z)^{c-a-b} F(c - b, c - a; 1 + c - a - b; 1 - z)$$
$$y_{18} = z^{1-c} (1 - z)^{c-a-b} F(1 - a, 1 - b; 1 + c - a - b; 1 - z)$$
$$y_{19} = z^{a-c}(1 - z)^{c-a-b} F\left(c - a, 1 - a; 1 + c - a - b; \frac{z - 1}{z}\right)$$
$$y_{20} = z^{b-c}(1 - z)^{c-b-1} F\left(1 - b, c - b; 1 + c - a - b; \frac{z - 1}{z}\right)$$
$$y_{21} = z^{1-c}(1 - z)^{c-a-b} F(1 - b, 1 - a; 2 - c; z)$$
$$y_{22} = z^{1-c} F(1 + b - c, 1 + a - c; 2 - c; z)$$
$$y_{23} = z^{1-c}(1 - z)^{c-b-1} F\left(1 - a, 1 + b - c; 2 - c; \frac{z}{z - 1}\right)$$
$$y_{24} = z^{1-c}(1 - z)^{c-a-1} F\left(1 + a - c, 1 - b; 2 - c; \frac{z}{z - 1}\right).$$

References

Askey R A 1975 Orthogonal Polynomials and Special Functions *Regional Conf. in Applied Math* vol 21 (Philadelphia, PA: SIAM)

Askey R A, Koonwinder T H and Shempp W 1984 *Special Functions: Group Theoretical Aspects and Applications* (Dordrecht: Reidelm)

Barnes E W 1908 *Proc. Land. Math. Soc.* **6** 141

Bell W W 1968 *Special Functions for Scientists and Engineers* (New York: von Nostrand)

Derbyshire J 2003 *Prime Obsession: Bernhard Riemann and the Greatest Unsolved Problem in Mathematics* (Washington, DC: John Henry Press)

Euler L 1748 *Introduction to Analysis Infinitorum* vol I (Lausanne: Bousquet)

Erdelyi A 1939 *Quart. J. Math., Oxford* **10** 129

Erdélyi A, Magnus W, Oberhettinger F and Tracomi F G 1953 *Higher Transcendental Functions* vols I, II (New York: McGraw Hill)

Gauss C F 1812 Disquisitiones generales circa serien infinitam *Thesis* Gottingen; *Ges. Werke Gottingen* **II** 437 **III**, 123, 207, 446.

Gauss C F 1876 *Werke* (Göttingen: Königlichen Gesellschaft der Wissenchaften)

Horner W G 1819 A new method of solving numerical equations of all orders by continuous approximation *Phil. Trans. R. Soc. Lond.* **109**

Kanigel R 1991 *The Man Who Knew Infinity* (New York: Washington Square Press)

Kummer E E 1975 *Collected Papers Vol. I: Contributions to Number Theory, Vol. II: Function Theory, Geometry and Miscellaneous* ed A Weil (Berlin: Springer)

Kummer E E 1836 Über die hypergeometrische Reihe $1 + \frac{\alpha\beta}{\gamma}\frac{x}{1!} + \frac{\alpha(\alpha+1)\beta(\beta+1)}{\gamma(\gamma+1)}\frac{x^2}{2!} + \ldots$ *J. für Math* **15** 39–83

Lakshminarayanan V and Varadarajan L 2015 *Special Functions for Optical Science and Engineering* (Bellingham, WA: SPIE Press)

Leavitt D 2008 *The Indian Clerk: A Novel* (NY: Bloomsbury)

Lievens S, Srinivasa Rao K and Vander Jeugt J 2005 *Integral Transform Spec. Funct.* **16** 153–8

Olver F W J, Lozier D W, Boisvert R F and Clark C W 2010 *NIST Handbook of Mathematical Functions* (Cambridge: Cambridge University Press)

Riemann B 1857 *Abh. d. Ges. d. Wiss. Gottinger.*

Slater Lucy J 1966 *Generalized Hypergeometric Functions* (Cambridge: Cambridge Univ. Press)

Srinivasa Rao K 1981 *Comp. Phys. Commun* **22** 297

Srinivasa Rao K and Vanden Berghe G 2004 *Historia Scientiarum* **13** 123

Srinivasa Rao K, Van der Jeugt J, Raynal J, Jagannathan R and Rajeswari V 1992 *J. Phys. A: Math. Gen.* **25** 861–76

Tent M B W 2008 *The Prince of Mathematics: Carl Friedrich Gauss* (Boca Raton, FL: CRC Press)

Wallis J 1655 *Arithmetica Infinitorum* (London: University of Oxford)

Weber Erfely M A 1952 *Am. Math. Monthly* **59** 163

IOP Publishing

Generalized Hypergeometric Functions
Transformations and group theoretical aspects
K Srinivasa Rao and Vasudevan Lakshminarayanan

Chapter 2

Group theory: basics

2.1 Introduction

There are three historical roots of group theory: the theory of algebraic equations, number theory, and geometry. Joseph-Louis Lagrange (1736–1813),[1] a great mathematician and physicist—known for his work in the areas of celestial mechanics, analytical mechanics, mathematical analysis, and number theory—is perhaps the earliest to study groups as such.

In brief, historically, a foundational root for the quest of the solutions to polynomial equations (of a degree greater than four) was the theory of groups. Lagrange, during 1770–71, laid the common foundation for the theory of equations on the basis of the group of permutations. Alexandre-Theophile Vandermonde (1735–96), a French mathematician, chemist, and a musician, is now principally associated with determinant theory. He studied symmetric functions and the solution of cyclotomic polynomials. He also studied a special class of matrices, which were later named after him, and in a contemporary paper anticipated the theory renowned today as Galois theory.

Vandermonde's identity is an elementary fact of combinatorics. Evariste Galois (1828; 1811–32) used permutation groups to describe how the various roots of a given polynomial equation are related to each other. Galois theory provides a connection between field theory and group theory. The modern approach to Galois theory, developed by Richard Dedekind (1831–1916) and Leopold Kronecker (1823–91); two German mathematicians, and Emil Artin (1898–1962); an Austrian mathematician, among others, involves studying the automorphisms of field extensions.

[1] He was actually born as Giuseppe Ludvico De la Grange in Turin, Italy. He wrote a famous four volume book on analytic classical mechanics, which was the foundation of mathematical physics in the nineteenth century. See this entry in Encyclopedia Britannica https://www.britannica.com/biography/Joseph-Louis-Lagrange-comte-de-lEmpire. See also Pepe (2014).

doi:10.1088/978-0-7503-1496-1ch2

Galois found that if r_1, r_2, ..., r_n are the n roots of an equation:

$$(x - r_1)(x - r_2)\cdots(x - r_n) = 0. \tag{2.1}$$

There is always a group of permutations of the roots r_i such that every function of the root's invariant to the substitutions of the group is rationally known. Conversely, every rationally determinable function of the root is invariant under the substitutions of the group.

Galois was the first to use the words *group* (*groupe*, in French) and *primitive* in their modern meanings. Galois introduced the concept of normal subgroups for a group. A normal subgroup is a subgroup that is invariant under conjugation in abstract algebra by members of the group of which it is a part. In other words, a subgroup H of a group G is normal if and only if $gH = Hg$ for all $g \varepsilon G$—i.e. the sets of left and right cosets, gH and Hg, respectively—coincide. Normal subgroups are used to construct quotient groups from a given group. A quotient group, or factor group, is a group obtained by aggregating similar elements of a larger group using an equivalence relation that preserves the group structure.

Galois discovered the notion of normal subgroups and found that a solvable primitive group may be identified to a subgroup of the affine group (a set of all non-singular affine transformations—i.e. transformations that preserve collinearity—of translations in a space, constituting the affine group) of an affine space over a finite field of prime order. Perhaps a complex set of reasons—such as his unhappy love affair, his failure to gain self-satisfying recognition for his mathematical work or political ambitions, his financial situation; all of which contributed to his weariness with life itself—led him to a duel with an alleged political friend, which resulted in his tragic death. Galois was mortally wounded in the duel, on May 30, 1832, at the age of twenty. He stayed up all night writing letters and composing the famous letter to Auguste Chevalier outlining his ideas, and attached three manuscripts as his mathematical testament. Hermann Weyl stated that 'this letter, if judged by the novelty and profundity of ideas it contains, is perhaps the most substantial piece of writing in the whole literature of mankind'. Before the duel, he wrote down in a long letter to his friend Auguste Chevalier (1873–1956) his mathematical theory and pointed out in it the cornerstones of his scientific life for a proper evaluation by posterity—see for instance, Rigatell (1996), Rothman (1982) or Infeld (1978).

Symmetry is one aspect of nature that man through the ages has admired and tried to comprehend to the best of his ability—symmetry is a vast subject. It is not only a fundamental concept but also an idea that unifies various branches of science. By the end of the 19th century, the mathematical notion of a group of transformations was available to the understanding of crystal structures. Around 100 years ago, the Russian crystallographer Federov (1971; 1853–1919) enumerated 17 two-dimensional symmetry groups—known as plane groups—that exhaust all possible planar symmetries with double infinite rapport. Rigorous proof for this was provided by George Polya (1974; 1887–1985) in 1924. Crystallographers established that there are only 32 point groups or 32 classes of crystals. In 1951, Shubnikov (Bradley and Cracknell 1972) introduced the notion of dichromatic symmetry—or, equivalently, anti-symmetry, or black-and-white symmetry—and provided a complete list of 1651 Shubnikov, or di-chromatic, space groups in three-dimensional space, which includes Federov's 17 monochromatic groups.

It is indispensable in studies of all areas of physics, ranging from elementary particle physics, condensed matter physics, optics, string theory, as well as in areas such as crystallography and spectroscopy. Symmetry studies also find applications in other diverse areas, such as image processing, statistics, and visual perception (see for example, Viana and Lakshminarayanan 2013, 2018).

In this chapter, we will give a brief overview of the basic concepts in group theory, necessary for understanding what follows in this book. For more information the reader should consult other texts, such as the ones by Zee (2013), Schwichtenberg (2015), Hamermesh (1962), Weyl (1939), or Naimark (1964).

2.2 Invariances, symmetries, and physics

Albert Einstein said 'the laws governing natural events are independent of the state of motion of the reference frame in which these events occur; if the frame travels without acceleration.' This principle is a cornerstone of Einstein's special theory of relativity, postulated by Einstein, in the miracle year (1905), according to which:

- all physical events take place in a four-dimensional space–time world (x, y, z, t), time being the fourth dimension;
- that all motion is relative motion;
- the velocity of light is the same for all observers regardless of their relative motion;
- the laws of physics are invariant under Lorentz transformations, which in the $x - t$ plane, for example, are:

$$x' = \frac{x - vt}{\sqrt{1 - v^2/c^2}}, \quad t' = \frac{t - xv/c}{\sqrt{1 - v^2/c^2}}, \quad y' = y, \quad z' = z \tag{2.2}$$

where v is the velocity of the moving frame of reference (x', y', z', t') with respect to the fixed frame (x, y, z, t) and c is the velocity of light.

Maxwell's universal laws for electromagnetism are Lorentz invariant[2]. Lorentz invariance is also referred to as relativistic invariance.

Invariance under translations and rotations means that an experiment carried out in one laboratory will give the same result in any other laboratory, and that the result would not depend on the direction in which the experiment is set up. Invariance under Lorentz transformations led to new formulations of both classical and quantum mechanics. However, relativistic corrections are important only at relativistic velocities; that is, velocities that are close to the velocity of light ($c = 2.997 \times 10^8\,\mathrm{ms}^{-1}$). Even then, they are small in most classical and quantum mechanical applications.

[2] Hendrik A Lorentz (1853–1928), Dutch physicist. For Lorentz invariance see Lorentz (1904) or Jackson (1975).

The Lorentz group that deals with rotations in a four-dimensional space–time world is a Lie group, which deals with the rotations in four-dimensional space (x, y, z, t), and is a continuous group with six parameters. The inclusion of space–time translations with the Lorentz group (consisting of rotations and boosts in four-dimensional space–time) generates the inhomogeneous Lorentz group, or the Poincaré group[3] (see Kim and Noz 2013).

In addition to relativistic invariance, all laws of physics must be invariant under Lorentz transformations, space–time translations, and three-dimensional rotations. These invariances lead to the dynamical laws of conservation of energy, momentum, angular momentum, and charge. The intimate connection between symmetries and conservation laws was first noticed in classical mechanics by Carl Gustar Jacob Jacobi (1884; 1804–51)[4] in 1842. Jacobi showed that for systems described by a classical Lagrangian, invariance of the Lagrangian under translations implies the conservation of linear momentum, and invariance of the Lagrangian under three-dimensional rotations implies the conservation of angular momentum.

In both classical and quantum mechanics, the conservation of linear momentum, angular momentum, and energy follows from the symmetries of the Hamiltonian under translations, rotations, and time translation. More generally, in quantum mechanics, whenever a conservation law holds for a physical system, the Hamiltonian of the system is invariant under a corresponding group of transformations. The converse of this statement is not true, which implies that even if the system has a Hamiltonian, which is invariant under a group of transformations, there may not be a corresponding conservation law—an example is invariance under time reversal. These realizations led to a powerful theorem called *Noether's theorem* (1918), named after Emmy Noether (1882–1935; see: James and Smith (1981); https://arstechnica.com/science/2015/05/the-female-mathematician-who-changed-the-course-of-physics-but-couldnt-get-a-job/):

> Whenever there is a symmetry in nature, there is also a conservation law and vice versa.

Or, conservation laws are necessary consequences of symmetries and symmetries necessarily entail conservation laws.

Noether's theorem is important, since it gives us insight into conservation laws. It is also a practical tool for calculations. Using this powerful theorem, we can determine from a physical system the conserved quantities (invariants) from the observed symmetries. We can also consider whole classes of hypothetical Lagrangians with given invariants to describe a physical system—that is, a physical theory that conserves a quantity X. By using this theorem, we can calculate the types of Lagrangians that conserve X through a continuous symmetry. The Lagrangian will provide the conditions and criteria necessary to understand the implications of the theory.

---

[3] Henri Poincaré, French mathematician (1854–1912). He was a polymath and also known as 'The Last Universalist' (Gray 2012). The Poinicaré group is used extensively in physics.

[4] A Lagrangian is a function that describes the state of a dynamic system in terms of generalized coordinates and is equal to the difference between potential and kinetic energy. See Lakshminarayanan *et al* (2001).

Noether's theorem is used in physics and also in the calculus of variations. This is a generalization of the formulation of the constants of motion used in Lagrangian and Hamiltonian formulations[5]. However, it does not apply to dissipative systems that have continuous symmetries and cannot be used to model systems that cannot be described by a Lagrangian alone. These implications are discussed elsewhere (Hanc *et al* 2004).

2.3 Discrete groups

Definition: a group G is a set of elements $S = \{g\}$ such that for any two elements g_1, $g_2 \,\varepsilon\, G$, a composition law $\circ$ called a product is defined. The group $(G, \circ)$ has four fundamental properties.

- Closue: if

$$g_1, g_2 \,\varepsilon\, G, \quad \text{then } g_1 \circ g_2 = g_3 \,\varepsilon\, G. \tag{2.3}$$

- Identity: there is an identity element $e \,\varepsilon\, G$ such that

$$g \circ e = e \circ g = g, \quad \text{for any } g \,\varepsilon\, G. \tag{2.4}$$

- Inverse: for every $g \,\varepsilon\, G$, there is an inverse $g^{-1} \,\varepsilon\, G$ such that

$$g \circ g^{-1} = g^{-1} \circ g = e. \tag{2.5}$$

  This inverse is unique. If $e = g^{-1} \circ g$ and also $e = \bar{g} \circ g$, then $g^{-1} \circ g = \bar{g} \circ g$. Right multiply by g^{-1}:

$$g^{-1} \circ g \circ g^{-1} = \bar{g} \circ g \circ g^{-1} \Rightarrow g^{-1} \circ e = \bar{g} \circ e \tag{2.6}$$

  therefore, $g^{-1} = \bar{g}$.
- Associativity: for every g_1, g_2, $g_3 \,\varepsilon\, G$,

$$g_1 \circ (g_2 \circ g_3) = (g_1 \circ g_2) \circ g_3. \tag{2.7}$$

Order. The order of a group is the number of elements in it. The order can be finite, denumerably infinite, i.e. the elements can be put in one-to-one correspondence with integers, or uncountably infinite.

Abelian. If all the elements of a group commute, then the group is said to be a commutative group or an Abelian group, named after the mathematician Abel[6]. It is an additive group, i.e. the group operation is an addition.

Some examples of finite groups are given below.

The simplest group is the trivial group consisting of only one element, the identity element e, since it is its own inverse.

---

[5] The Hamiltionian is an operator corresponding to the total energy of a system. Named after the Irish theoretical physicist William Rowan Hamilton (1805–1865). see Hankins (2004).

[6] Niels Henrik Abel (1802–29), Norwegian mathematician. See Ore (2008) and http://www.brittania.com/biography/Niels-Henrik-Abel.

The next simplest group consists of two elements, the two-element groups, with the identity, e, and the other element, a, with the multiplication or product law:

$$e \cdot e = e, \quad e \cdot a = a = a \cdot e, \quad a \cdot a = e, \text{ and } a^{-1} = a, \tag{2.8}$$

where, since $a \cdot a \neq a^2$, we must have $a^2 = e$, i.e. a is its own inverse. The law of combination is multiplication. Note that {0,1}, is a two-element group with the law of combination being addition modulo 2:

$$0 + 0 = 0, \quad 0 + 1 = 1, \quad 1 + 0 = 1, \quad 1 + 1 = 0. \tag{2.9}$$

The following are examples of two element groups:

- $\{e = 0, a = 1\}$ with addition modulo 2, as the law of combination:

$$0 + 0 = 0, \quad 0 + 1 = 1, \quad 1 + 0 = 1, \quad 1 + 1 = (1)0, \tag{2.10}$$

  with (1) the carry digit, which is the binary law of addition, the law used in the design of the adder, the building block of the modern digital computer;
- $\{e = 1, a = -1\}$, with multiplication as the law of combination. The group associated with the numbers ± 1 is important in quantum mechanics because it can be associated with charge conjugation (C) and two other discrete transformations for space reflection (or, parity P) and time reversal invariance (T). Together, the product of these three transformations give rise to a fundamental theorem called the CPT theorem[7]. This theorem states that CPT symmetry holds for all physical phenomena and that any Lorenz invariant local quantum theory with a Hermitian Hamiltonian must have CPT symmetry. See Streater and Wightman (1964);
- $e = I$, $(x, y, z = 0)$; $a = (-x, y, z)$, reflection in YZ-plane;
- $e = I$, $(x = 0, y = 0, z = 0)$; $a = (-x, -y, -z)$, reflection about the origin;
- same as above, with a rotation through 180° about the Z-axis;
- a is the reciprocal transformation in the unit sphere. The point with spherical coordinates (r, θ, ϕ) is imaged in the point $(\frac{1}{r}, \theta, \phi)$;
- The elements are permutations of two symbols:

$$e = \begin{Bmatrix} 1 \rightarrow 1 \\ 2 \rightarrow 2 \end{Bmatrix}; \quad a = \begin{Bmatrix} 1 \rightarrow 2 \\ 2 \rightarrow 1 \end{Bmatrix} = P_{12}. \tag{2.11}$$

The following are also some examples of groups.

1. ..., −3, −2, −1, 0, 1, 2, 3,... is the group of integers $(\mathcal{Z}, +)$.
2. p/q, the rationals, $\mathcal{Q}$, including 0, under addition.
3. The complex numbers $(\mathcal{C},+)$.
4. Even integers, $2n$, $\forall n = 0, 1, 2, \infty$, (2Z,+).

[7] CPT—charge, parity, and time reversal symmetry—theorem states that CPT symmetry holds for all physical phenomena, and that any Lorentz invariant local quantum theory with a Hermitian Hamiltonian must have CPT symmetry. See Greenberg (2006).

Note. Two group G, G' are isomorphic: $G \sim G'$ if their elements can be put int one-to-one correspondence and this correspondence is preserved under the law of composition. (1) and (4) are isomorphic for infinite order only.

5. Powers of $\ldots 2^{-2}, 2^{-1}, 2^0, 2^1, 2^2, \ldots,$ under multiplication. (1) and (5) are isomorphic for infinite order.
6. $\mathcal{Q}$, excluding zero, $(\mathcal{Q}, \times)$.
7. $\mathcal{R}$, real numbers, excluding zero, $(\mathcal{R}/0,\times)$.
8. $\mathcal{C}$, group of complex numbers, excluding zero, $(\mathcal{C}/0,\times)$.
9. $\exp 2\pi m\ i\ n$, $m = 0, 1, 2, \ldots,$ nth roots of unity, is a cyclic group of finite order, under multiplication.
10. Non-singular $n \times n$ matrices, with matrix multiplication.
11. The unimodular group of $n \times n$ matrices, with det $= \pm 1$, with matrix multiplication.
12. The special unimodular $n \times n$ matrices, with det $= +1$, with matrix multiplication.

All the above examples are discrete groups of infinite order.

2.4 The symmetric group S_n

The symmetric group of order n, has as elements permutation of degree n, of n objects. An element of this group S_n is denoted by:

$$\begin{pmatrix} 1 & 2 & \cdots & n \\ s_1 & s_2 & \cdots & s_n \end{pmatrix} \tag{2.12}$$

The number of elements, or the order of the group is $n!$ For, s_1 in the above element of the group can be represented by any one of the n symbols: $1,2, \ldots, n$. After this is done, s_2 can be replaced by anyone of the remaining $(n-1)$ symbols; s_3 by any one of the remaining $(n-2)$ symbols; until s_n can be replaced by the last of the symbols not assigned. Thus, the order of the group S_n is: $n(n-1)(n-2)\cdots 2 \cdot 1 = n!$, satisfying the properties:

1. Closure:

$$\begin{pmatrix} 1 & 2 & \cdots & n \\ s_1 & s_2 & \cdots & s_n \end{pmatrix}\begin{pmatrix} 1 & 2 & \cdots & n \\ t_1 & t_2 & \cdots & t_n \end{pmatrix} = \begin{pmatrix} 1 & 2 & \cdots & n \\ t_{s1} & t_{s2} & \cdots & t_{sn} \end{pmatrix} \varepsilon\ S_n \tag{2.13}$$

2. Identity:

$$\begin{pmatrix} 1 & 2 & \cdots & n \\ 1 & 2 & \cdots & n \end{pmatrix} \tag{2.14}$$

3. Inverse:

$$\begin{pmatrix} 1 & 2 & \cdots & n \\ s_1 & s_2 & \cdots & s_n \end{pmatrix}^{-1} = \begin{pmatrix} s_1 & s_2 & \cdots & s_n \\ 1 & 2 & \cdots & n \end{pmatrix} \tag{2.15}$$

4. Associativity:

$$\begin{aligned}
&\begin{pmatrix} 1 & 2 & \cdots & n \\ u_1 & u_2 & \cdots & u_n \end{pmatrix}\left[\begin{pmatrix} 1 & 2 & \cdots & n \\ t_1 & t_2 & \cdots & t_n \end{pmatrix}\begin{pmatrix} 1 & 2 & \cdots & n \\ s_1 & s_2 & \cdots & s_n \end{pmatrix}\right] \\
&\quad = \begin{pmatrix} 1 & 2 & \cdots & n \\ u_1 & u_2 & \cdots & u_n \end{pmatrix}\begin{pmatrix} 1 & 2 & \cdots & n \\ s_{t1} & s_{t2} & \cdots & s_{tn} \end{pmatrix} \\
&\quad = \begin{pmatrix} 1 & 2 & \cdots & n \\ s_{t_{u1}} & s_{t_{u2}} & \cdots & s_{t_{un}} \end{pmatrix} = LHS.
\end{aligned} \tag{2.16}$$

$$\begin{aligned}
&\left[\begin{pmatrix} 1 & 2 & \cdots & n \\ u_1 & u_2 & \cdots & u_n \end{pmatrix}\begin{pmatrix} 1 & 2 & \cdots & n \\ t_1 & t_2 & \cdots & t_n \end{pmatrix}\right]\begin{pmatrix} 1 & 2 & \cdots & n \\ s_1 & s_2 & \cdots & s_n \end{pmatrix} \\
&\quad = \begin{pmatrix} 1 & 2 & \cdots & n \\ t_{u1} & t_{u2} & \cdots & t_{un} \end{pmatrix}\begin{pmatrix} 1 & 2 & \cdots & n \\ s_1 & s_2 & \cdots & s_n \end{pmatrix} \\
&\quad = \begin{pmatrix} 1 & 2 & \cdots & n \\ s_{t_{u1}} & s_{t_{u2}} & \cdots & s_{t_{un}} \end{pmatrix} = RHS.
\end{aligned} \tag{2.17}$$

LHS = RHS and hence the proof of the associativity property of S_n.

As an example, for illustration, consider the permutation group S_3. The order of the group is 3! = 6. Let the six elements of the group be:

$$\begin{aligned}
&e = \begin{pmatrix} 1 & 2 & 3 \\ 1 & 2 & 3 \end{pmatrix}; \quad a = \begin{pmatrix} 1 & 2 & 3 \\ 2 & 1 & 3 \end{pmatrix} = P_{12}; \quad b = \begin{pmatrix} 1 & 2 & 3 \\ 1 & 3 & 2 \end{pmatrix} = P_{23}; \\
&c = \begin{pmatrix} 1 & 2 & 3 \\ 3 & 2 & 1 \end{pmatrix} = P_{13}; \quad d = \begin{pmatrix} 1 & 2 & 3 \\ 3 & 1 & 2 \end{pmatrix}; \quad f = \begin{pmatrix} 1 & 2 & 3 \\ 2 & 3 & 1 \end{pmatrix}.
\end{aligned} \tag{2.18}$$

The product of any two elements of S_3, $a \circ b$ (say) is:

$$\begin{aligned}
a \circ b = P_{12}P_{23} &= \begin{pmatrix} 1 & 2 & 3 \\ 2 & 1 & 3 \end{pmatrix}\begin{pmatrix} 1 & 2 & 3 \\ 1 & 3 & 2 \end{pmatrix} \\
&= \begin{pmatrix} 1 & 3 & 2 \\ 2 & 3 & 1 \end{pmatrix}\begin{pmatrix} 1 & 2 & 3 \\ 1 & 3 & 2 \end{pmatrix} = \begin{pmatrix} 1 & 2 & 3 \\ 2 & 3 & 1 \end{pmatrix} = f.
\end{aligned} \tag{2.19}$$

Note that in the last line we used a property of the elements of the permutation group:

$$\begin{pmatrix} 1 & 2 & 3 \\ 2 & 1 & 3 \end{pmatrix} = \begin{pmatrix} 1 & 3 & 2 \\ 2 & 3 & 1 \end{pmatrix}, \tag{2.20}$$

where a column-interchange has been made to make the implementation of the product rule transparent.

The elements e, d and f constitute the elements of the cyclic group C_3 of three elements. It is a subgroup of S_3. For

$$\begin{aligned} d \circ f &= \begin{pmatrix} 1 & 2 & 3 \\ 3 & 1 & 2 \end{pmatrix} \begin{pmatrix} 1 & 2 & 3 \\ 2 & 3 & 1 \end{pmatrix} \\ &= \begin{pmatrix} 2 & 3 & 1 \\ 1 & 2 & 3 \end{pmatrix} \begin{pmatrix} 1 & 2 & 3 \\ 2 & 3 & 1 \end{pmatrix} \\ &= \begin{pmatrix} 1 & 2 & 3 \\ 1 & 2 & 3 \end{pmatrix} = e, \text{ the identity.} \end{aligned} \tag{2.21}$$

Consider a permutation of degree 8:

$$\begin{pmatrix} 1 & 2 & 3 & 4 & 5 & 6 & 7 & 8 \\ 2 & 3 & 1 & 5 & 4 & 7 & 6 & 8 \end{pmatrix} \equiv (2\,3\,1\,5\,4\,7\,6\,8), \tag{2.22}$$

referred to as an eight-cycle. Starting with symbol 1, we see that the permutation takes $1 \to 2$, $2 \to 3$, $3 \to 1$, closing a *cycle* (123). Similarly, $4 \to 5$, $5 \to 4$ closes the cycle (45); (67) and 8 are the other cycles. The eight-cycle permutation can be written in terms of these as:

$$(2\,3\,1\,5\,4\,7\,6\,8) = (123)(45)(67)(8) \equiv (123)(45)(67). \tag{2.23}$$

A cycle has no letters/symbols in common. Thus the eight-cycle can written as:

$$\begin{pmatrix} 1 & 2 & 3 & 4 & 5 & 6 & 7 & 8 \\ 2 & 3 & 1 & 4 & 5 & 6 & 7 & 8 \end{pmatrix} \equiv (1\,2\,3). \tag{2.24}$$

Since they have no common symbols, cycles can be permuted:

$$(123)(45) = (45)(123). \tag{2.25}$$

Individual cycles can start at any point in the chain, so that:

$$(123) = (231) = (312). \tag{2.26}$$

A cycle with two symbols, a two-cycle, is called a *transposition*.

Any cycle can be written uniquely as a product of transpositions (having elements in common). Thus,

$$(123) = \begin{pmatrix} 1 & 2 & 3 \\ 2 & 3 & 1 \end{pmatrix} = \begin{pmatrix} 2 & 1 & 3 \\ 2 & 3 & 1 \end{pmatrix} \begin{pmatrix} 1 & 2 & 3 \\ 2 & 1 & 3 \end{pmatrix} = (13)(12), \tag{2.27}$$

and, in general:

$$(12 \cdots n) = (1n) \cdots (13)(12). \tag{2.28}$$

Or, a three-cycle, is a product of two transpositions, and an n-cycle is a product of $(n - 1)$ transpositions. Thus, the above example for an eight-cycle can be written as:

$$(23154768) = (123)(45)(67) = (13)(12)(45)(67). \tag{2.29}$$

The number of permuted symbols is seven and the number of independent cycles is three. Their difference, $7 - 4 = 3$, is called the *decrement* of the permutation. If the decrement is even(odd), its resolution into a product of transpositions will have an even(odd) number of factors. Permutations with even(odd) decrements are said to be even(odd) permutations.

2.5 An interesting property of S_n

A new property of the symmetric group, S_n, was recently reported by Srinivasa Rao and Pandir (2016). It arises when each element of S_n is assigned a unique place value, enabling an ordering of the elements (numerically). They showed that the differences between successive elements of this ordered sequence give rise to a palindromic sequence $\mathcal{P}_n$. Sieving out a given element at a time in the maximal palindromic sequence of the group, S_n, of length $n! - 1$, results in a hierarchy of palindromic sequences, ending with a single element, which will be the central element of $\mathcal{P}_n$. The consequences of the concept of the place value ordering the elements of S_n were studied and are summarized in this section.

Let $<$ be a lexicographic ordering on the elements of S_n. Consider the two permutations of S_n, $\sigma = (a_1\, a_2 \cdots a_n)$ and $\pi = (b_1\, b_2 \cdots b_n)$. We say $\sigma < \pi$, if there exists $i \in [n]$: $\forall j < i$, $\sigma_{(j)} = \pi_{(i)}$ and $\sigma_{(i)} < \pi(j)$, where $\sigma_{(i)}$ and $\pi_{(j)}$ denote the ith and jth element in permutation σ and π, respectively, and $[n] \equiv \{1, 2, \ldots, n\}$.

Consider the smallest non-trivial symmetric group S_3, which is a set of six elements (2.18). Assigning the lexicographic ordering on the permutations of $\{1, 2, 3\}$ leads to the 'standard' ordering e, b, a, f, d, c, which is explicitly

$$123 \to 132 \to 213 \to 231 \to 312 \to 321. \tag{2.30}$$

The following is a definition for assigning place-value to the permutations belonging to S_n:

$$f_B : S_n \to \mathcal{N}, \quad f_B(\sigma) = \Sigma_{i=1}^{n} \sigma(i)\, B^{n-i}, \quad \forall \quad B \in N, \quad B \geqslant 2. \tag{2.31}$$

f_B will be referred to as the *place-valued* function. This definition enables (for the first time) *arithmetic operations* to be made on the permutations. Differences between successive/adjacent place-value ordered permutations lead to a sequence of numbers. For instance, in the case of S_3 (2.18), with $B = 10$, the five differences between the six specifically ordered permutations give rise to the sequence:

$$9, 81, 18, 81, 9, \tag{2.32}$$

which is a *palindromic* sequence.

Let $S_n = \{\sigma_1, \sigma_2, \ldots, \sigma_{n!}\}$ denote the place-valued, lexicographically ordered symmetric group. The palindromic sequence

$$(\mathcal{P}_{\|})^{\beta}_{\infty \leqslant \| \leqslant /!-\infty}, \tag{2.33}$$

on a particular value of $B = \beta$, is associated with $\mathcal{S}_{l!}$, given by the elements

$$\mathcal{P}_k = \sum_{1 \leqslant i \leqslant n} (b_i - a_i) \times \beta^{n-i}, \tag{2.34}$$

where

$$a_i \in \sigma_j \quad \text{and} \quad b_i \in \sigma_{j+1}, \quad \forall \quad 1 \leqslant j \leqslant (n! - 1). \tag{2.35}$$

The following are essential definitions required to prove the main theorem:

1. Let $S_n = \{\sigma_1, \sigma_2, \ldots, \sigma_{n!}\}$ denote the place-valued, lexicographically ordered symmetric group.
2. The palindromic sequence $(\mathcal{P}_{||})^{\beta}_{\infty \leqslant || \leqslant l! - \infty}$, on a particular value of $B = \beta$, is associated with $\mathcal{S}_{l!}$, given by the elements

$$\mathcal{P}_k = \sum_{1 \leqslant i \leqslant n} (b_i - a_i) \times \beta^{n-i}, \tag{2.34}$$

where

$$a_i \in \sigma_j \quad \text{and} \quad b_i \in \sigma_{j+1}, \quad \forall \quad 1 \leqslant j \leqslant (n! - 1). \tag{2.35}$$

The following is the statement of the main theorem:

Theorem. *The differences between a successive, place-value assigned, lexicographically ordered sequence of elements of the symmetric group* $(\mathcal{S})_{(l!)}$, *generating a palindromic sequence,* $(\mathcal{P})^{\beta}_{1 \leqslant k \leqslant n! - 1}$, *of length n! − 1.*

The palindromic sequence $(\mathcal{P}_n)_{n!-1}$, associated with S_n, is given by the elements.

Before we give the proof of the theorem, we introduce some definitions.

Consider the set of permutations,

$$(\mathcal{S})_{(n!)} = \{\sigma_1, \sigma_2, \ldots, \sigma_{(n!)}\} \, \varepsilon \, S_n. \tag{2.38}$$

Given a permutation $\sigma = (a_1, a_2, \ldots, a_n)$, consider the function

$$f : \mathcal{S}(n!) \to \mathcal{S}(n!), \tag{2.39}$$

where

$$f(\sigma) = f(a_1\, a_2 \cdots a_n) = h(a_1)\, h(a_2) \cdots h(a_n). \tag{2.40}$$

Consider the function

$$h : \{1\, 2 \cdots n\} \to \{1\, 2 \cdots n\}, \tag{2.41}$$

where

$$\begin{aligned} &h(1) \to n,\ h(2) \to n - 1,\ \ldots h(i) \\ &\quad \to n - i + 1,\ \cdots h(n-1) \to 2,\ h(n) \to 1. \end{aligned} \tag{2.42}$$

The following claim gives insight, over the mapping done by function f, among the elements of the set $\mathcal{S}(n!)$:

$$f(\sigma_j) = \sigma_{n!-j+1}. \tag{2.43}$$

Proof. Let $P(j)$ be the statement $f(\sigma_j) = \sigma_{n!-j+1}$. We present the proof by induction on the index of permutations in $\mathcal{S}(n!)$.

The basic step required is

$$\begin{aligned} f(\sigma_1) &= f(1\ 2\ 3 \ldots n-1\ n) = f(1)f(2)\ldots f(n-1)f(n) \\ &= n(n-1)\ldots 3\ 2\ 1 = \sigma_{n!}. \end{aligned} \tag{2.44}$$

The next is an inductive step:

Let $P(j)$ be true for all j with $2 \leqslant j \leqslant k-1$. Consider

$$\sigma_{k-1} = (s_1, s_2, \ldots, s_{m-1}, s_m, s_{m+1}, \ldots, s_n). \tag{2.45}$$

The next permutation in lexicographic order, is

$$\sigma_k = (s_1', s_2', \ldots, s_{m-1}', s_m', s_{m+1}' \ldots, s_n'). \tag{2.46}$$

Since σ_{k-1} and σ_k are adjacent permutations, $s_i = s_i'$, for all $1 \leqslant i \leqslant (m-1)$, it follows that $h(s_i) = h(s_i')$. Now s_m' is the smallest element amongst $(s_{m+1}, s_{m+2}, \ldots, s_n)$, such that $s_m' > s_m$, so that $h(s_m') < h(s_m)$, and since

$$s_{m+1}' < s_{m+2}' < \ldots s_n', \quad \Rightarrow \quad h(s_{m+1}') > h(s_{m+2}') > \ldots > h(s_n'), \tag{2.47}$$

and $h(\sigma_k)$ gives just the permutation, prior to $h(\sigma_{k-1})$ and by the induction hypothesis

$$h(\sigma_{k-1}) = \sigma_{n!-(k-1)+1} \quad \Rightarrow \quad h(\sigma_k) = \sigma_{n!-k+1}. \tag{2.48}$$

Hence the proof.

Consider two permutations

$$\sigma_p = (a_1, a_2, \ldots, a_n), \qquad \sigma_q = (b_1, b_2, \ldots, b_n) \tag{2.49}$$

for all $p < q \leqslant n!$, from the left end of the set $\mathcal{S}(n!)$. Let

$$\begin{aligned} f(\sigma_p) &= \sigma_{n!-p+1} = (d_1, d_2, \ldots, d_n), \\ f(\sigma_q) &= \sigma_{n!-q+1} = (c_1, c_2, \ldots, c_n) \end{aligned} \tag{2.50}$$

be the pth and the qth elements of the same set, but from the right end.

Lemma:

$$(b_i - a_i) = (d_i - c_i), \quad \forall\, 1 \leqslant i \leqslant n. \tag{2.51}$$

Proof: by claim 1,

$$\begin{aligned} f(\sigma_p) &= f(a_1, a_2, \ldots, a_n) = h(a_1)h(a_2)\ldots h(a_n) \\ &= (d_1\ d_2 \ldots d_n) = \sigma_{n!-p+1}. \end{aligned} \tag{2.52}$$

Similarly,

$$\begin{aligned} f(\sigma_q) &= f(b_1, b_2, \dots, b_n) = h(b_1)h(b_2) \dots h(b_n) \\ &= (c_1\, c_2 \dots c_n) = \sigma_{n!-q+1}. \end{aligned} \tag{2.53}$$

Note that

$$n! - p + 1 > n! - q + 1 \;\forall\; p < q \leqslant n!. \tag{2.54}$$

Let $a_i = \ell$ and $b_i = r$ where $\ell, r \in \{1, 2, 3, \dots n\}$ are two arbitrary elements in σ_p and σ_q, respectively. Now

$$\begin{aligned} d_i - c_i &= h(a_i) - h(b_i) = (n - \ell + 1) - (n - r + 1) \\ &= r - \ell = b_i - a_i. \end{aligned} \tag{2.55}$$

Hence the proof.

Now we prove our main theorem.

Proof. The difference between the decimal place values of the permutations σ_j and σ_{j+1}, from the left end of S, can be given as:

$$\sum_{1 \leqslant i \leqslant n} (b_i - a_i) \times B^{n-i}. \tag{2.56}$$

Similarly, the difference between the decimal place values of the permutations $\sigma_{n!-j+1}$ and $\sigma_{n!-(j+1)+1}$, from the right end of $\mathcal{S}(n!)$, can be given as:

$$\sum_{1 \leqslant i \leqslant n} (d_i - c_i) \times B^{n-i} \equiv \sum_{1 \leqslant i \leqslant n} (b_i - a_i) \times B^{n-i}, \tag{2.57}$$

since by lemma 1,

$$(b_i - a_i) = (d_i - c_i) \quad \forall 1 \leqslant i \leqslant n. \tag{2.58}$$

So we get the same number from both ends when we take the difference between two successive permutations of $\mathcal{S}(n!)$, indices at j and $(j + 1)$, and consequently the resultant sequence of length $(n! - 1)$, is *palindromic*. Hence the proof.

2.6 Representations of a group

A *representation* of a group is a homomorphism (or, mapping) of a group G into a group of linear operators acting on a linear vector space. If the linear operators are matrices, the representation is called a matrix representation. Unless otherwise specified, a representation is always a matrix representation.

Matrices occur naturally in transformations, as in the case of:

$$x_i' = \sum_j \alpha_{ij} x_j, \quad (i, j = 1, 2, 3), \tag{2.59}$$

where α is a 3×3 matrix, which satisfies the group properties:

1. Identity:

$$\alpha_{ij} = \delta_{ij} \Rightarrow \text{unit matrix I}. \tag{2.60}$$

2. Closure:

$$\sum_j \alpha_{ij} \alpha_{jk} = \alpha_{ik}. \tag{2.61}$$

3. Inverse: if α is non-singular, its inverse α^{-1} exists, such that

$$\alpha\alpha^{-1} = I = \alpha^{-1}\alpha. \tag{2.62}$$

4. Associativity:

$$\sum_{j,k} \alpha_{ij}\alpha_{jk}\alpha_{k\ell} = \sum_k \left(\sum_j \alpha_{ij}\alpha_{jk} \right) \alpha_{k\ell} = \alpha_{i\ell} \tag{2.63}$$

$$= \sum_j \alpha_{ij} \left(\sum_k \alpha_{jk}\alpha_{k\ell} \right) = \sum_j \alpha_{ij}\alpha_{j\ell} = \alpha_{i\ell} \tag{2.64}$$

and equation (2.63) = equation (2.64), implying associativity.

If $\det \alpha = \pm 1$, then the representation is said to be *unimodular*. The subset of transformations having $\det \alpha = +1$ constitute a *subgroup* H of the group G. If to every member g_i of the group G, we associate a matrix

$$D(g_i) : g_i \rightarrow D(g_i), \tag{2.65}$$

then, for the matrices to form a representation of G, the mapping must preserve the law of composition, which is matrix multiplication.

$$g_i \leftrightarrow D(g_i), \quad g_j \leftrightarrow D(g_j), \quad g_i \circ g_j \leftrightarrow D(g_i)D(g_j), \forall g_i, g_j \, \varepsilon \, G. \tag{2.66}$$

The following example provides a representation for the permutation group of three elements, S_3:

$$e \Rightarrow D(e) = \begin{pmatrix} 1 & 0 & 0 \\ 0 & 1 & 0 \\ 0 & 0 & 1 \end{pmatrix}, \qquad a \Rightarrow D(12) = \begin{pmatrix} 0 & 1 & 0 \\ 1 & 0 & 0 \\ 0 & 0 & 1 \end{pmatrix},$$
$$b \Rightarrow D(23) = \begin{pmatrix} 1 & 0 & 0 \\ 0 & 0 & 1 \\ 0 & 1 & 0 \end{pmatrix}, \qquad c \Rightarrow D(13) = \begin{pmatrix} 0 & 0 & 1 \\ 0 & 1 & 0 \\ 1 & 0 & 0 \end{pmatrix}, \tag{2.67}$$
$$d \Rightarrow D(312) = \begin{pmatrix} 0 & 1 & 0 \\ 0 & 0 & 1 \\ 1 & 0 & 0 \end{pmatrix}, \qquad f \Rightarrow D(231) = \begin{pmatrix} 0 & 0 & 1 \\ 1 & 0 & 0 \\ 0 & 1 & 0 \end{pmatrix}.$$

Corresponding to the product of two permutations:

$$(12)(23) = \begin{pmatrix} 1 & 2 & 3 \\ 2 & 1 & 3 \end{pmatrix}\begin{pmatrix} 1 & 2 & 3 \\ 1 & 3 & 2 \end{pmatrix} = \begin{pmatrix} 1 & 2 & 3 \\ 2 & 3 & 1 \end{pmatrix} = (231), \tag{2.68}$$

the matrix representation is:

$$D(12)D(23) = \begin{pmatrix} 0 & 1 & 0 \\ 1 & 0 & 0 \\ 0 & 0 & 1 \end{pmatrix}\begin{pmatrix} 1 & 0 & 0 \\ 0 & 0 & 1 \\ 0 & 1 & 0 \end{pmatrix} = \begin{pmatrix} 0 & 0 & 1 \\ 1 & 0 & 0 \\ 0 & 1 & 0 \end{pmatrix} = (231). \tag{2.69}$$

If a representation is *isomorphic* to the group, then it is said to be a *faithful* representation of the group. All groups of matrices are faithful representations of themselves.

Two representations, D_1 and D_2, are *equivalent* if they are related to each other by a *similarity transformation*:

$$D_2(g) = S\, D_1(g)\, S^{-1}, \quad \forall\, g\, \varepsilon\, G, \tag{2.70}$$

with a fixed operator S. This transformation is called a similarity transformation.

A representation D is *reducible* if it is equivalent to a representation D' with *block diagonal* form:

$$D'(g) = S\, D(g)\, S^{-1} = \begin{pmatrix} D_1'(g) & 0 \\ 0 & D_2'(g) \end{pmatrix}. \tag{2.71}$$

The linear vector space on which D' acts breaks up into two orthogonal subspaces, each of which is mapped into itself. The representation D' is said to be a *direct sum*:

$$D' = D_1' + D_2'. \tag{2.72}$$

A representation is *irreducible* if it has been reduced to block diagonal form and cannot be reduced any further by a similarity transformation.

2.7 Lie groups and Lie algebras

'I am certain, absolutely certain that these theories will be recognized as fundamental at some point in the future', prophesied their inventor, Sophus Lie (1842–99). Lie groups and Lie algebras are an integral part of the tools of the mathematician and physicist today. Sophus Lie was born, in 1842 in Western Norway. He was enrolled at the Royal Fredrik University, Oslo, in 1859, and graduated in 1865. Disappointed at being second in his class, he left the university. To get over his depression, he went on long hiking tours. Ludvig Sylow, one of his teachers at the Frederik University, never guessed that Lie would realize, by 1868, that 'there was a mathematician hidden inside him'. Lie started with Euclid's renowned book, *Elements*, and continued with books, describing non-Euclidean geometry, a subject which was in its infancy at that time.

In December 1869, his first scientific publication (Lie 1869) won him a travel scholarship, which enabled him to make his first tour abroad, during the course of which he became one of the most renowned mathematicians in Europe.

Lie's doctoral dissertation resounded through Europe's mathematical communities. A year later he was appointed as a professor in Christania, where he remained for almost 14 years. Lie's lectures were incredibly inspiring and extremely enlightening to his students. In 1886, Lie moved to Leipzig as a professor with his wife and three children. Unfortunately, his 'beautiful mind' led him to a state of paranoia, and a fear of a lack of peer recognition. This led to his nervous breakdown and, in turn, to his accusing other mathematicians of stealing his ideas, without referring to his works. Though Lie wanted to be back in Norway, it took several years before he was offered a suitable position in Christania, in 1898. On 18 February, 1899, Lie died of pernicious anaemia (see Stubhaug and Daly 2002 for a fascinating biography).

> Lie theory is in the process of becoming the most important part of modern mathematics. Little by little it became obvious that the most unexpected theories, from arithmetic to quantum physics, came to encircle this Lie field like a gigantic axis, said the french scholar Jean Dieudonne.

In contrast to finite groups, Lie groups have an infinite number of elements, and so are characterized by the number of continuous parameters. For example, rotations in two dimensions:

$$x = r\cos\theta, \qquad y = r\sin\theta \tag{2.73}$$

are characterized by the two continuous parameters (r, θ), where r is the radius of a circle and $0 \leqslant \theta \leqslant 2\pi$ is the angle of rotation, a continuous parameter. Rotations in three dimensions are characterized in the spherical polar coordinate transformations:

$$x = r\sin\theta\cos\phi, \quad y = r\sin\theta\sin\phi, \quad z = r\cos\theta \tag{2.74}$$

with r as the radius of the sphere, $0 \leqslant \theta \leqslant \pi$ and $0 \leqslant \phi \leqslant 2\pi$.

With each Lie group is associated a Lie algebra. We consider a typical group of fundamental importance and relevance in physics; the rotation group in three dimensions, characterized by the orthogonal transformations (2.74), and hence called an orthogonal group, $O(3)$. Good expositions of Lie groups and Lie algebras can be found in Lipkin (2002) and Gilmore (2006).

2.8 Angular momentum and the rotation group

A particle of mass m and velocity $\vec{v}$, located at a point $\vec{r}$, measured from the origin of a coordinate system, has a *linear momentum* $\vec{p} = m\vec{v}$ and an *angular momentum*:

$$\vec{L} = \vec{r} \times \vec{p}. \tag{2.75}$$

The classical mechanical concept becomes an operator in quantum mechanics, since momentum $\vec{p}$ is replaced by $-i\hbar\nabla$, $\hbar = \mathrm{h}/2\pi$; h being the Planck's constant, $i = \sqrt{-1}$ and ∇ the gradient operator, which in the orthogonal Cartesian coordinate system is:

$$\nabla = \hat{i}\frac{\partial}{\partial x} + \hat{j}\frac{\partial}{\partial y} + \hat{k}\frac{\partial}{\partial z} \tag{2.76}$$

where $\hat{i}$, $\hat{j}$, $\hat{k}$ are the unit vectors along the orthogonal Cartesian coordinate axes X, Y, Z, respectively.

Note: it is often convenient and customary to use the *Natural System of Units* in which $\hbar = c = m = 1$, or the Compton wavelength of a particle of mass m, $\hbar/(mc) = 1$ is set as the unit of length throughout the calculations/computations and, in the final step, the result of the calculation/computation is multiplied by the required M, L, T (mass, length and time) dimensional units. Accordingly,

$$\begin{aligned} L_x &= y\,p_z - z\,p_y = -i\left(y\frac{\partial}{\partial z} - z\frac{\partial}{\partial y}\right), \\ L_y &= z\,p_x - x\,p_z = -i\left(z\frac{\partial}{\partial x} - x\frac{\partial}{\partial z}\right), \\ L_z &= x\,p_y - y\,p_x = -i\left(x\frac{\partial}{\partial y} - y\frac{\partial}{\partial x}\right). \end{aligned} \tag{2.77}$$

The commutator $[A, B]$ of the two operators A, B is defined as:

$$[A, B] = A\,B - B\,A, \tag{2.78}$$

which plays a pivotal role in quantum mechanics. According to quantum mechanics, the components of the momentum operator are:

$$p_x = -i\frac{\partial}{\partial x}, \quad p_y = -i\frac{\partial}{\partial y}, \quad p_z = -i\frac{\partial}{\partial z}. \tag{2.79}$$

The component of the position operator of a particle and its corresponding momentum satisfies the commutation relations (in natural units):

$$[x, p_x] = [y, p_y] = [z, p_z] = i. \tag{2.80}$$

For,

$$
\begin{aligned}
[x, p_x]\,\psi(x) &= (x\, p_x - p_x\, x)\,\psi(x) \\
&= -\, i\left(x\frac{\partial}{\partial x} - \frac{\partial}{\partial x}x\right)\psi(x) \\
&= -\, i\, x\frac{\partial\psi(x)}{\partial x} + i\frac{\partial}{\partial x}(x\,\psi(x)) \\
&= -\, i\, x\frac{\partial\psi(x)}{\partial x} + i\,\psi(x) + i\, x\frac{\partial\psi(x)}{\partial x}
\end{aligned}
\tag{2.81}
$$

and with the cancellation of the $i\, x\frac{\partial}{\partial x}\psi(x)$ term, we get the required result:

$$
[x, p_x]\,\psi(x) = i\,\psi(x) \Rightarrow [x, p_x] = i. \tag{2.82}
$$

All other commutation relations between the components of $\vec{r}$ and $\vec{p}$ are zero, i.e.,

$$
[x_k, p_l] = i\,\delta_{k\ell}, \tag{2.83}
$$

where the Kronecker delta function $\delta k\ell$ is defined as:

$$
\delta_{k\ell} = \begin{cases} +\,1, \text{ for k} = \ell \\ 0, \text{ for k} \neq \ell. \end{cases} \tag{2.84}
$$

Using this commutation relation, which is Heisenberg's uncertainty principle (1902–76),[8] we can derive the commutation relations satisfied by the components of orbital angular momentum.

Explicitly, it follows that the position vector of a particle and its momentum satisfy the basic commutation relations:

$$
[x, p_x] = i, \quad [x, p_y] = 0 = [x, p_z], \quad \text{cyclically.} \tag{2.85}
$$

The commutation relations of the Cartesian components of orbital angular momentum $\vec{L}$ can also be readily derived as follows:

$$
\begin{aligned}
[L_x, L_y] &= [y\, p_z - z\, p_y, z\, p_x - x\, p_z] \\
&= [y\, p_z, z\, p_x] - [y\, p_z, x\, p_z] - [z\, p_y, z\, p_x] + [z\, p_y, x\, p_z]
\end{aligned}
\tag{2.86}
$$

where the required commutators $[A\ B, C\ D]$ are derived as follows:

$$
\begin{aligned}
[A\,B, C\,D] &= [X, C\,D], \qquad \text{where } X = A\,B \quad \text{(say)} \\
&= X\,C\,D - C\,D\,X \\
&= X\,C\,D - C\,X\,D + C\,X\,D - C\,D\,X \\
&= [X, C]\,D + C\,[X, D]
\end{aligned}
\tag{2.87}
$$

[8] Werner Karl Heisenberg (1901–76) is a German physicist. One of the original founders of quantum mechanics. See Cassidy (1991) and Mott and Peierls (1977).

$$
\begin{aligned}
[A\,B,\ Y] &= A\,B\,Y - Y\,A\,B \\
&= A\,B\,Y - A\,Y\,B + A\,Y\,B - Y\,A\,B \\
&= A\,[B,\ Y] + [A,\ Y]\,B.
\end{aligned} \tag{2.88}
$$

Therefore,

$$
\begin{aligned}
[AB,\ CD] &= (A[B,\ C] + [A,\ C]B)\,D + C(A[B,\ D] + [A,\ D]B) \\
&= A\,[B,\ C]\,D + [A,\ C]\,BD + CA\,[B,\ D] + C\,[A,\ D]\,B.
\end{aligned} \tag{2.89}
$$

Hence,

$$[y\,p_z,\ z\,p_x] = y[p_z,\ z]\,p_x = -i\,y\,p_x \tag{2.90}$$

$$[y\,p_z,\ x\,p_z] = 0 = [z\,p_y,\ z\,p_x] \tag{2.91}$$

$$[z\,p_y,\ x\,p_z] = x\,[z,\ p_z]\,p_y = i(x\,p_y - y\,p_x) = i\,L_z. \tag{2.92}$$

Similarly,

$$[L_y,\ L_z] = i\,L_x \tag{2.93}$$

$$[L_z,\ L_x] = i\,L_y. \tag{2.94}$$

The three commutation relations for the components of $\vec{L}$ can be written in terms of the Levi-Civita tensor as:

$$[L_k,\ L_\ell] = i\,\varepsilon_{k\ell m}\,L_m, \tag{2.95}$$

where

$$
\varepsilon_{k\ell m} = \begin{cases} +\,1, & \text{for even permutations of (123)} \\ -\,1, & \text{for odd permutations of (123)} \\ 0, & \text{otherwise.} \end{cases} \tag{2.96}
$$

The square of the angular momentum operator is:

$$L^2 = L_x^2 + L_y^2 + L_z^2 \tag{2.97}$$

and it commutes with all the components of $\vec{L}$, so that

$$[L^2,\ L_k] = 0, \quad k = (x,\ y,\ z). \tag{2.98}$$

Quantum states can be specified by the simultaneous eigenfunctions of L^2 and L_z or, any one component of ($\vec{L}$). These (L^2, L_z) constitute the *complete set of commuting generators* for orbital angular momentum. If another component of $\vec{L}$ is included in this set, it will not commute with L_z. The measurement of another variable corresponding to an operator not commuting with the set L^2 L_z, necessarily introduces uncertainty into one of the variables already measured. A sharper specification of the system is therefore not possible.

A general angular momentum operator $\vec{J}$ is defined as one whose Cartesian components obey the commutation relations:

$$[J_k, J_\ell] = i\, \varepsilon_{k\ell m}\, J_m. \tag{2.99}$$

This extended definition permits the existence of spin—a quantity that has no classical analog.

In the case of orbital angular momentum $\vec{L}$, it is well known from the study of partial differential equations that the solution of the Laplace's equation is:

$$\nabla^2 \psi = 0. \tag{2.100}$$

In spherical polar coordinates, the solution by the method of separation of variables can be written as:

$$\psi(r, \theta, \phi) = R(r)\, Y_m^\ell(\theta, \phi) \tag{2.101}$$

where $Y_m^\ell(\theta, \phi)$ satisfies the differential equation:

$$\left[\frac{1}{\sin\theta}\frac{d}{d\theta}\left(\sin\theta\frac{d}{d\theta}\right) + \ell(\ell+1) - \frac{m^2}{\sin^2\theta}\right]\Theta_{\ell,\,m}(\theta) = 0 \tag{2.102}$$

and

$$\left[\frac{d^2}{d\phi^2} + m^2\right]\Phi_m(\phi) = 0 \tag{2.103}$$

with

$$Y_m^\ell(\theta, \phi) = \Theta_{\ell,\,m}(\theta)\, \Phi_m(\phi), \tag{2.104}$$

called the spherical harmonics (see Lakshminarayan and Varadharajan 2015, pp 221–41).

In this representation, in analogy with the eigenvalue problem for orbital angular momentum, viz:

$$L^2\, Y_m^\ell(\theta, \phi) = \ell(\ell+1)\, \hbar^2\, Y_m^\ell(\theta, \phi) \tag{2.105}$$

$$L_z\, Y_m^\ell(\theta, \phi) = m\, \hbar\, Y_m^\ell(\theta, \phi), \tag{2.106}$$

which are also written in the Dirac[9] bra-ket notation, in which the spherical harmonics satisfy the eigenvalue equations:

$$Y^{\ell}_{m}(\theta, \phi) = \Theta_{\ell, m}(\theta)\, \Phi(\phi) \equiv |\ell, m\rangle \tag{2.107}$$

$$L^2\, |\ell, m\rangle = \ell(\ell + 1)\, |\ell, m\rangle \tag{2.108}$$

$$L_z\, |\ell, m\rangle = m\, |\ell, m\rangle. \tag{2.109}$$

For general angular momentum—which can be either orbital ($\vec{L}$), spin ($\vec{\mathbf{S}}$), or their sum $\vec{J} = \vec{L} + \vec{S}$, called total angular momentum—one constructs the eigenstates of the general angular momentum in the bra-ket (< *bra*|*c*|*ket* >) notation of Dirac. The simultaneous eigenfunctions of J^2, J_z is:

$$J^2\, |j, m\rangle = j(j + 1)\, |j, m\rangle \tag{2.110}$$

$$J_z\, |j, m\rangle = m\, |j, m\rangle \tag{2.111}$$

in the natural units $\hbar = c = 1$.

Note: in these equations, m stands for the projection of the angular momentum $\vec{J}$ and not mass (as in $E = mc^2$).

The operator

$$J_x^2 + J_y^2 + J_z^2 - J_z^2 \tag{2.112}$$

is diagonal in $|j, m\rangle$ representation and it has a positive-definite (non-negative) eigenvalue:

$$(J_x^2 + J_y^2)\, |j, m\rangle = (J^2 - J_z^2)\, |j, m\rangle = (\lambda_j - m^2)\, |j, m\rangle, \tag{2.113}$$

because the expectation value of the square of a Hermitian operator, i.e. the square of a real eigenvalue, is greater than or equal to zero. Therefore,

$$\lambda_j = j(j + 1)\, \hbar^2 = < \langle j, m|\, J^2\, |j, m\rangle, \quad \text{and} \quad \lambda_j \geqslant m^2. \tag{2.114}$$

Thus, the value of m is bounded by both above and below and m^2 cannot exceed λ_j. This implies that for a given $\vec{J}$, there exist minimum and maximum values of m, denoted by m_{min} and m_{max}.

Let $J_\pm$ define the raising and lowering operators:

$$J_\pm = J_x \pm i\, J_y, \tag{2.115}$$

[9] Paul Adrian Maunce Dirac (1902–84), British physicist and Nobelist. See Farmelo (2011).

which satisfy the commutation relations:

$$[J^2, J_\pm] = 0 = [J^2, J_z] \tag{2.116}$$

and

$$[J_z, J_\pm] = \pm J_\pm \tag{2.117}$$

$$[J_+, J_-] = 2 J_z. \tag{2.118}$$

Now examine the behaviour of the function $J_\pm |j, m\rangle$:

$$J^2 J_\pm |j, m\rangle = J_\pm J^2 |j, m\rangle = \lambda_j J_\pm |j, m\rangle \tag{2.119}$$

$$J_z J_\pm |j, m\rangle = (J_\pm J_z \pm J_\pm) |j, m\rangle = (m \pm 1) J_\pm |j, m\rangle. \tag{2.120}$$

From these, it follows that $J_\pm |j, m\rangle$ is an eigenfunction of J^2 with an eigenvalue λ_j and, also, an eigenfunction of J_z with the eigenvalue $(m \pm 1)$.

Thus, $J_\pm |j, m\rangle$ is proportional to the normalized eigenfunction $|j, m \pm 1\rangle$, i.e.

$$J_\pm |j, m\rangle = C_\pm |j, m \pm 1\rangle, \tag{2.121}$$

where $C_\pm$ is a proportionality constant. The ability of the operators $J_\pm$ to raise/lower the value of m by ± 1 unit, respectively, while preserving λ_j, gives them their nomenclature as raising/lowering, step-up/step-down, and ladder or shift operators.

Since the values of m are bounded between m_{min} and m_{max}, it follows that

$$J_+ |j, m_{\text{max}}\rangle = 0 = J_- |j, m_{\text{min}}\rangle. \tag{2.122}$$

By applying J_- to the first and J_+ to the second and using:

$$J^2 = J_x^2 + J_y^2 + J_z^2 = \frac{1}{2}(J_+J_- + J_-J_+) + J_z^2 \tag{2.123}$$

$$= \frac{1}{2}(J_+J_- + J_+J_- - 2J_z) + J_z^2, \quad \text{since } [J_+, J_-] = 2J_z \tag{2.124}$$

$$J^2 = J_+J_- + J_z(J_z - 1), \tag{2.125}$$

so that

$$J_\mp J_\pm = J^2 - J_z(J_z + 1), \tag{2.126}$$

we obtain

$$\begin{aligned} J_-J_+ |j, m_{\text{max}}\rangle &= (J^2 - J_z(J_z + 1)) |j, m_{\text{max}}\rangle \\ &= \lambda_j - m_{\text{max}}(m_{\text{max}} + 1) = 0 \end{aligned} \tag{2.127}$$

and

$$\begin{aligned} J_+J_- \,|j, m_{\min}\rangle &= (J^2 - J_z(J_z - 1))\,|j, m_{\min}\rangle \\ &= \lambda_j - m_{\min}(m_{\min} - 1) = 0. \end{aligned} \tag{2.128}$$

Eliminating λ_j from these two equations:

$$m_{\max}(m_{\max} + 1) = m_{\min}(m_{\min} - 1) \tag{2.129}$$

or

$$(m_{\max} + m_{\min})(m_{\max} - m_{\min} + 1) = 0 \tag{2.130}$$

and one of the two factors must vanish. Since $m_{\max} \geqslant m_{\min}$, the only possible solution is:

$$m_{\max} = -m_{\min}. \tag{2.131}$$

From $J_\pm \,|j, m\rangle = C_\pm \,|j, m \pm 1\rangle$, we also know that successive eigenvalues of m differ by their unity. Therefore, $m_{\max} - m_{\min}$ is a positive definite integer, which we denote by 2j, where j is an *integer* or *half-integer*, i.e.

$$m_{\max} - m_{\min} = 2j \qquad \text{and} \qquad m_{\max} + m_{\min} = 0, \tag{2.132}$$

so that $m_{\max} = j$, and $m_{\min} = -j$. Or,

$$-j \leqslant m \leqslant j \tag{2.133}$$

$$i.\,e. \quad m = -j, -j + 1, -j + 2, \ldots, -1, 0, 1, 2, \ldots, j - 2, j - 1, j. \tag{2.134}$$

Thus,

$$\lambda_j = m_{\max}(m_{\max} + 1) = m_{\min}(m_{\min} - 1)\, j(j + 1) = -j(-j - 1). \tag{2.135}$$

We are now in a position to evaluate the proportionality constant, $C_\pm$:

$$\begin{aligned} \langle j, m|\, J_\mp J_\pm \,|j, m\rangle &= \langle j, m|\, J^2 - J_z(J_z \pm 1)\,|j, m\rangle \\ &= j(j + 1) - m(m \pm 1) = (j \mp m)(j \pm m + 1). \end{aligned} \tag{2.136}$$

Note: in the theory of angular momentum the replacement:

$$j \to -j - 1 \tag{2.137}$$

is a trivial mathematical symmetry.

Furthermore,

$$\langle j, m|\, J_\mp J_\pm \,|j, m\rangle = \langle j, m|\, J_\mp \sum_{j',m'} |j', m'\rangle\langle j', m'|\, J_\pm \,|j, m\rangle, \tag{2.138}$$

using the completeness property of the quantum mechanical states:

$$\sum_{\nu} |\nu\rangle\langle\nu| = 1, \tag{2.139}$$

we have

$$\begin{aligned}\langle j, m| J_{\mp}J_{\pm} |J_{\pm} |j, m\rangle &= \langle j, m| J_{\mp} |j, m \pm 1\rangle\langle j, m \pm 1| J_{\pm} |j, m\rangle \\ &= (\langle j, m \pm 1| J_{\pm} |j, m\rangle)^{\dagger}\langle j, m \pm 1| J_{\pm} |j, m\rangle \\ &= C_{\pm}^{\dagger}C_{\pm} = |C_{\pm}|^2.\end{aligned} \tag{2.140}$$

The absolute value of $C_{\pm}$ is determined up to an arbitrary phase. The Condon and Shortley convention adopts the positive root. So,

$$C_{\pm} = [j(j+1) - m(m \pm 1)]^{1/2} = [(j \mp m)(j \pm m + 1)]^{1/2}, \tag{2.141}$$

i.e., $C_{\pm}$ is real and the matrix elements of J_x are real while those of J_y are purely imaginary:

$$\begin{aligned}\langle j', m'| J^2 |j, m\rangle &= j\,(j+1)\,\delta_{jj'}\,\delta_{mm'} \\ \langle j', m'| J_z |j, m\rangle &= m\,\delta_{jj'}\,\delta_{mm'} \\ \langle j', m'| J_{\pm} |j, m\rangle &= [(j \mp m)(j \pm m + 1)]^{1/2}\,\delta_{j'j}\,\delta_{m', m\pm 1}.\end{aligned} \tag{2.142}$$

The last of the above is equivalent to:

$$\langle j', m'| J_x |j, m\rangle = \frac{1}{2}[(j \mp m)(j \pm m + 1)]^{1/2}\,\delta_{j'j}\,\delta_{m', m\pm 1} \tag{2.142a}$$

$$\langle j', m'| J_y |j, m\rangle = \mp\frac{i}{2}[(j \mp m)(j \pm m + 1)]^{1/2}\,\delta_{j'j}\,\delta_{m', m\pm 1} \tag{2.142b}$$

where $\delta_{m,n}$ is the Kronecker delta function:

$$\delta_{m,n} = \begin{cases} 0 & m \neq n \\ 1 & m = n. \end{cases} \tag{2.143}$$

As an exercise let us find the two-dimensional (and three-dimensional) representations of the angular momentum operators:

$\langle m'|J_z|m\rangle$: (2.144a)

m'/m	$+\frac{1}{2}$	$-\frac{1}{2}$
$+\frac{1}{2}$	$\frac{1}{2}$	0
$-\frac{1}{2}$	0	$-\frac{1}{2}$

	$m' = m + 1/m$	$+\frac{1}{2}$	$-\frac{1}{2}$
$\langle m'\vert J_+\vert m\rangle$:	$+\frac{1}{2}$	0	1
	$-\frac{1}{2}$	0	0

(2.144*b*)

and

	$m' = m - 1/m$	$+\frac{1}{2}$	$-\frac{1}{2}$
$\langle m'\vert J_-\vert m\rangle$:	$+\frac{1}{2}$	0	0 .
	$-\frac{1}{2}$	1	0

(2.144*c*)

Therefore,

$$J_z = 1/2\begin{pmatrix} 1 & 0 \\ 0 & -1 \end{pmatrix}; \quad J_+ = 1/2\begin{pmatrix} 0 & 1 \\ 0 & 0 \end{pmatrix}; \quad J_- = 1/2\begin{pmatrix} 0 & 0 \\ 1 & 0 \end{pmatrix}. \tag{2.145}$$

Since $J_\pm = J_x \pm iJ_y$

$$J_x = \frac{1}{2}(J_+ + J_-) = \frac{1}{2}\begin{pmatrix} 0 & 1 \\ 1 & 0 \end{pmatrix} \tag{2.146a}$$

$$J_y = \frac{1}{2i}(J_+ - J_-) = \frac{1}{2i}\begin{pmatrix} 0 & 1 \\ -1 & 0 \end{pmatrix} = \frac{1}{2}\begin{pmatrix} 0 & -i \\ i & 0 \end{pmatrix}. \tag{2.146b}$$

Note: when $J_k = \frac{1}{2}\sigma_k$

$$\sigma_z = \begin{pmatrix} 1 & 0 \\ 0 & -1 \end{pmatrix}, \quad \sigma_x = \begin{pmatrix} 0 & 1 \\ 1 & 0 \end{pmatrix}, \quad \sigma_y = \begin{pmatrix} 0 & -i \\ i & 0 \end{pmatrix} \tag{2.147}$$

the σ_k, $(k = 1, 2, 3)$ matrices are the Pauli[10] spin-1/2 matrices.

The corresponding spin 1 matrices are:

	m'/m	$+1$	0	-1
$\langle m'\vert J_z\vert m\rangle$:	$+1$	1	0	0
	0	0	0	0
	-1	0	0	-1

(2.148*a*)

[10] Named after Wolfgang Ernst Pauli 1900–58, Swiss physicist and Nobelist. See library.com/archives/Pauli-archive/biography. See also Enz (2010).

$$\langle m+1|J_+|m\rangle: \quad \begin{array}{c|ccc} m+1/m & +1 & 0 & -1 \\ \hline +1 & 0 & 1 & 0 \\ 0 & 0 & 0 & 1 \\ -1 & 0 & 0 & 0 \end{array} \tag{2.148b}$$

and

$$\langle m-1|J_-|m\rangle: \quad \begin{array}{c|ccc} m-1/m & +1 & 0 & -1 \\ \hline +1 & 0 & 0 & 0 \\ 0 & 1 & 0 & 0 \\ -1 & 0 & 1 & 0 \end{array}. \tag{2.148c}$$

Therefore,

$$J_z = \begin{pmatrix} 1 & 0 & 0 \\ 0 & 0 & 0 \\ 0 & 0 & -1 \end{pmatrix}; \quad J_+ = \begin{pmatrix} 0 & 1 & 0 \\ 0 & 0 & 1 \\ 0 & 0 & 0 \end{pmatrix}; \quad J_- = \begin{pmatrix} 0 & 0 & 0 \\ 1 & 0 & 0 \\ 0 & 1 & 0 \end{pmatrix}. \tag{2.149}$$

Since $J_\pm = J_x \pm iJ_y$

$$J_x = \frac{1}{2}(J_+ + J_-) = \frac{1}{2}\begin{pmatrix} 0 & 1 & 0 \\ 1 & 0 & 1 \\ 0 & 1 & 0 \end{pmatrix} \tag{2.150a}$$

$$J_y = \frac{1}{2i}(J_+ - J_-) = \frac{1}{2}\begin{pmatrix} 0 & -i & 0 \\ i & 0 & -i \\ 0 & i & 0 \end{pmatrix}. \tag{2.150b}$$

The operator $J^2 = J_x^2 + J_y^2 + J_z^2$ is a *Casimir operator,* which commutes with all other generators:

$$[J^2, J_i] = 0. \tag{2.151}$$

The angular momentum commutation relations:

$$[J_a, J_b] = i\,\varepsilon_{abc}\,J_c, \qquad (a, b, c) \equiv (x, y, z) \equiv (1, 2, 3) \tag{2.152}$$

represent the simplest *non-Abelian Lie algebra,* with ε_{abc} being the *structure constants.* The summation convention implies that there is a sum over c, but in this case, it is trivial due to the properties of the Levi-Civita tensor and the commutator $[J_1, J_2]$, giving the 'other' (third) generator J_3. e.g. $[J_1, J_2] = i\,J_3$, and it is a *Lie algebra.*

Note: $J_i = \frac{1}{2}\sigma_i$, $i = 1, 2, 3$ the generators provide a suitable (but not unique) representation of the Pauli matrices σ_i, which have the properties:

$$\sigma_i \, \sigma_j = \delta_{ij} + i \, \varepsilon_{ijk} \, J_k, \tag{2.153}$$

which implies $\sigma_i^2 = 1$.

Irreducible representations are characterized by the highest $J_z \equiv J_3$ eigenvalue j, which must be an integer or half-odd integer, and they are $(2j + 1)$-dimensional. The eigenvalues of J_3 are called *weights* and j is the highest weight.

The construction for angular momentum—characterizing the irreducible representations (Irreps or IRRs) in terms of their highest weight—generalizes any Lie algebra. The representation with the highest weight j is sometimes called the *spin j representation.*

When Hermitian angular momentum operators exist, the entire Hilbert[11] space is a representation and can be decomposed into states labeled $|\alpha \; j \; m\rangle$ in the spin j representation with $J_z = m$. The α label stands for all other labels necessary to characterize the states. The state can be chosen to be orthonormal in α:

$$\langle \alpha' \, j' \, m' | \, \alpha \, j \, m\rangle = \delta_{\alpha,\,\beta} \, \delta_{j,\,j'} \, \delta_{m,\,m'}. \tag{2.154}$$

These states form a *basis* for studying the properties of the system under rotations.

Just as any representation can be broken down into simpler irreducible representations, an arbitrary representation cab be built out of the simplest representation. The simplest non-trivial representation is generated by the Pauli matrices:

$$J_a = \sigma_a/2; \quad \sigma_1 = \begin{pmatrix} 0 & 1 \\ 1 & 0 \end{pmatrix}; \quad \sigma_2 = \begin{pmatrix} 0 & -i \\ i & 0 \end{pmatrix}; \quad \sigma_3 = \begin{pmatrix} 1 & 0 \\ 0 & -1 \end{pmatrix}. \tag{2.155}$$

The group elements $\exp[i\lambda_a\sigma_a/2]$ are the special uintary 2×2 matrices, and the group is called $SU(2)$, which stands for special unitary unimodular group.

In mathematics, the special unitary group of degree n, denoted by $SU(n)$, is the Lie group of $n \times n$ unitary matrices with the determinant 1. The group operation is matrix multiplication. The special unitary group is a subgroup of the unitary group $U(n)$, consisting of all $n \times n$ unitary matrices. As a compact classical group, $U(n)$ is the group that preserves the standard inner product on $\mathcal{C}^n$. It is itself a subgroup of the general linear group,

$$SU(n) \subset U(n) \subset GL(n, \mathcal{C}). \tag{2.156}$$

The $SU(n)$ groups find wide application in the standard model of particle physics, especially $SU(2)$ in the electroweak interaction and $SU(3)$ in quantum chromo dynamics.

The simplest case, $SU(1)$, is the trivial group, having only a single element. The group $SU(2)$ is isomorphic to the group of quaternions of the norm 1, and is thus diffeomorphic to the three-sphere. Since unit quaternions can be used to represent rotations in three-dimensional space (up to sign), there is a surjective

[11] Named after David Hilbert (1862–1943), German mathematician. See Reid (1996).

homomorphism from $SU(2)$ to the rotation group $SO(3)$. $SU(2)$ is also identical to one of the symmetry groups of spinors, $Spin(3)$, which enables a spinor representation of rotations.

Given two representations, $D_1(g)$ and $D_2(g)$, of a group of dimensions, m and n, respectively, it is possible to form another representation of dimension $m + n$ as their *direct sum*, which is formed by the block diagonal matrices

$$\begin{bmatrix} D_1(g) & 0 \\ 0 & D_2(g) \end{bmatrix}. \tag{2.157}$$

One can also form a representation of dimension $n \times m$ as follows: if $|i\rangle$, $i = 1, 2, \dots, n$ is an orthonormal basis on which D_1 acts and $|x\rangle$, $x = 1, 2, \dots, n$ is an orthonormal basis on which D_2 acts, then the product $|i\rangle\,|x\rangle$ is an orthonormal basis in an $n \times m$ dimensional space called the *direct product* space. In this space, the direct product is:

$$(D_1 \otimes D_2)(g)\{|i\rangle|x\rangle\} = D_1(g)\,|i\rangle\, D_2(g)\,|x\rangle. \tag{2.158}$$

In matrix language:

$$\langle y|\,\langle j|\,(D_1 \otimes D_2)(g)\,|i\rangle|x\rangle = \langle j|D_1(g)\,|i\rangle\,\langle j|\,D_2(g)\,|x\rangle, \tag{2.159a}$$

$$i.\,e.\ [(D_1(g) \otimes D_2(g)]_{ij,\,xy} = [D_1(g)]_{ij}\,[D_2(g)]_{xy}. \tag{2.159b}$$

We can form a direct product of n two-dimensional representations as

$$[D \otimes \cdots \otimes D]_{i_1 i_2 \cdots i_n,\, j_1 \cdots j_n} = D_{i_1 j_1} \cdots D_{i_n j_n} \tag{2.160}$$

acting on the n-component objects $U_{j_1 \cdots j_n}$. This representation is reducible.

The rotation group: consider $\mathcal{R}^3$, the real three-dimensional vector space:

$$\mathcal{R}^3 = \{\vec{x} = (x_1, x_2, x_3 \mid x_i \text{ real})\}. \tag{2.161}$$

The linear transformations:

$$\vec{x} \to R\,\vec{x} \qquad \text{or} \qquad x' = R_{ij}x_j, \quad i, j = 1, 2, 3, \tag{2.162}$$

which preserve the scalar product:

$$(x', y') = \sum_i x_i\, y_i \tag{2.163}$$

are such that:

$$(x', y') = R_{ij}\, x_j\, R_{ik}\, y_k = R_{ij}\, R_{ik}\, x_j\, y_k \equiv (x, y), \tag{2.164}$$

which implies

$$\sum_{j,k} R_{ij}\, R_{ik} = \delta_{j,\,k} \qquad \text{or} \qquad R\,R^T = I, \tag{2.165}$$

where I is the 3×3 unit matrix. Hence, R is real and orthogonal.

The set of all real, orthogonal matrices with $\det R = +1$ form the subgroup $SO(3)$ of special orthogonal transformations in real three-dimensional space. They do not include inversions—or space reflections. The set of all matrices with $\det R = -1$ do not form a subgroup, since the identity does not belong to this subset, which is not closed under multiplication.

Let the element

$$P = \begin{pmatrix} -1 & 0 & 0 \\ 0 & -1 & 0 \\ 0 & 0 & -1 \end{pmatrix} \varepsilon\ O(3) \tag{2.166}$$

be the parity operator with

$$\det P = -1 \qquad \text{and} \qquad P^2 = I. \tag{2.167}$$

If $R\ \varepsilon\ O(3)$ and $\det R = -1$, let

$$R = P\,R', \quad \text{where} \quad R' = PR. \tag{2.168}$$

Now $R'\ \varepsilon\ O(3)$ and $\det R' = \det(PR) = +1$, which implies that $R'\ \varepsilon\ SO(3)$. Therefore,

$$O(3) = SO(3) \cup PSO(3). \tag{2.169}$$

The corresponding groups in the n-dimensions are $O(n)$ and $SO(n)$, with

$$O(n) = SO(n) \cup PSO(n) \tag{2.170}$$

where $P = -I$, I being the 3×3 unit matrix.

The group $SU(2)$ consists of 2×2 unitary, unimodular matrices in two-dimensional complex vector space, which leaves

$$(z, z') = \sum_i z_i^* z_i' \quad \text{invariant.} \tag{2.171}$$

A similar definition holds for $SU(n)$ where $g\ \varepsilon\ SU(n)$ is an $n \times n$ unitary, unimodular matrix.

The Lorentz group: consider the transformation in four-dimensional space–time:

$$x_m u' = \sum_{\nu=0}^{3} \Lambda_{\mu\nu}\, x_\nu, \qquad \mu, \nu = 0, 1, 2, 3 \tag{2.172}$$

where Λ is a real 4×4 matrix, which leaves as invariant the scalar product

$$(x, y) = x_0\, y_0 - x_i\, y_i, \tag{2.173}$$

using the Einstein summation convention, according to which repeated indices imply a summation over that index for the range specified. In the four dimensions, the range of index is 0,1,2,3. The same is written succinctly as:

$$(x, y) = x_\mu g_{\mu\nu}\, x_\nu \qquad \text{and} \qquad \Lambda^T g\, \Lambda = g, \tag{2.174}$$

where

$$g_{\mu\nu} = \begin{cases} +1 & \text{for} \quad \mu, \nu = 0 \\ -1 \ \text{for} & \mu, \nu = 1, 2, 3 \end{cases} \tag{2.175}$$

is called the *metric tensor*. In matrix form g, there will be a diagonal matrix with the diagonal elements being

$$g_{00} = 1, g_{11} = g_{22} = g_{33} = -1. \tag{2.176}$$

The Poincaré group[12], P, consists of the Lorentz transformations and four translations along the four axes. An element of P is written as (a, Λ), where $a = (a_0, a_1, a_2, a_3)$ and the transformations are:

$$x'_\mu = \Lambda_{\mu\nu}\, x_\nu + a_\mu, \text{ where } \Lambda \,\varepsilon\, \mathcal{L},\ a \,\varepsilon\, \mathcal{R}^4 \tag{2.177}$$

and $\mathcal{L}$ represents the Lorentz group. The Poincaré transformation is also written as

$$x' = (a, \Lambda)\, x = \Lambda x + a, \tag{2.178}$$

i.e. first the Lorentz transform and then translate—this choice of order is a convention.

The group properties for the Poincaré transformations are:

(i) Closure:

$$\begin{aligned} &(a', \Lambda')(a, \Lambda)\, x = (a'\Lambda')(\Lambda x + a) \\ &= \Lambda'(\Lambda x + a) + a' = \Lambda'\Lambda x + \Lambda' a + a' = (\Lambda' a + a', \Lambda'\Lambda)\, x. \end{aligned} \tag{2.179}$$

(ii) Identity:

$$(a, \Lambda) = (0, I), I \text{ being the } 4 \times 4 \text{ unit matrix.} \tag{2.180}$$

(iii) Inverse:

$$(a, \Lambda)^{-1}(a, \Lambda)(a, \Lambda)\, x = (a, \Lambda)^{-1}(\Lambda x + a) \tag{2.181}$$

$$= (-\Lambda^{-1}a, \Lambda^{-1})(\Lambda x + a) = \Lambda^{-1}(\Lambda x + a) - \Lambda^{-1}a = x. \tag{2.182}$$

Therefore,

$$(a, \Lambda)^{-1} = (-\Lambda^{-1}a, \Lambda^{-1}). \tag{2.183}$$

(iv) Associativity:

$$(a'', \Lambda'')((a', \Lambda')(a, \Lambda)) = (a'', \Lambda'')(\Lambda' a + a', \Lambda'\Lambda) \text{ using } (i) \tag{2.184}$$

$$= (\Lambda''(\Lambda' a + a') + a'', \Lambda''\Lambda'\Lambda) = (\Lambda''\Lambda' a + \Lambda'' a' + a'', \Lambda''\Lambda'\Lambda) \tag{2.185}$$

---

[12] See Gray (2002).

and

$$((a'', \Lambda'')(a', \Lambda'))(a, \Lambda) = (\Lambda'' a' + a'', \Lambda''\Lambda')(a, \Lambda) \tag{2.186}$$

$$= (\Lambda''\Lambda' a + \Lambda'' a' + a'', \Lambda''\Lambda'\Lambda) \tag{2.187}$$

so that

$$(a'', \Lambda'')((a', \Lambda')(a, \Lambda)) = ((a'', \Lambda'')(a', \Lambda'))(a, \Lambda). \tag{2.188}$$

2.9 Compact Lie groups

Compact Lie groups are groups of unitary operators in which the group elements are labeled by a set of continuous parameters with a multiplicative law that depends smoothly on the parameters. *Compactness* is a global property.

Any representation of a compact Lie group is equivalent to a representation of unitary operators. Any group element that can be obtained from the identity by continuous changes in the parameters can be written as: $\exp[i\alpha_a x_a]$, where $\alpha_a(a = 1, 2, \ldots, N)$ are real parameters and X_a are linear independent Hermitian operators. The set of all linear combinations $\alpha_a X_a$ are called the *generators* of the group or as *basis vectors*. They form a linear vector space. They satisfy the simple commutation relations, which determine the (almost) full structure of the group.

If $\exp[i\lambda X_a]$ and $\exp[i\lambda X_b]$ are elements in a commutative (abelian) group, then

$$\exp[i\lambda X_b]\ \exp[i\lambda X_a]\ \exp[i\lambda X_b] = \exp[i\lambda X_a]. \tag{2.189}$$

But if the group is non-commutative, then

$$\exp[i\lambda X_b]\ \exp[i\lambda X_a]\ \exp[i\lambda X_b]\ \exp[-i\lambda X_a] = \gamma = \exp[i\beta_c X_c] \tag{2.190}$$

measures the amount by which

$$\exp[i\lambda X_b]\ \exp[i\lambda X_a]\ \exp[i\lambda X_b]\ \text{differesfrom}\ \exp[i\lambda X_a] \tag{2.191}$$

and γ is a group element. For, close to the identity:

$$\begin{aligned} &(1 + i\lambda X_b)(1 + i\lambda X_a)(1 + i\lambda X_b)(1 - i\lambda X_a) \\ &= 1 + \lambda^2[X_a, X_b] + \lambda^2(X_a^2 + X_b^2) \\ &\equiv \exp[i\beta_c X_c] = 1 + i\beta_c X_c, \end{aligned} \tag{2.192}$$

since the generators are infinitesimal, as $\lambda \to 0$, we must have:

$$\lambda^2[X_a, X_b] \to i\beta_c X_c. \tag{2.193}$$

Let $\beta_c = \lambda^2 f_{abc}$, so that:

$$[X_a, X_b] = i f_{abc} X_c, \tag{2.194}$$

where f_{abc} are called *structure constants* determined by the group multiplication law.

The properties of the Lie group near the identity element are called infinitesimal properties. Sophus Lie established the following theorems for constructing the Lie Algebra, associated with any Lie group:

Theorem. The structure constants are constant. If X_a are the generators of a Lie group, then

$$[X_a, X_b] = i f_{abc} X_c, \tag{2.194}$$

where f_{abc} are the structure constants.

Theorem. The generators obey the *Jacobi identity*:

$$[X_a, [X_b, X_c]] + \text{cyclic permutations} = 0. \tag{2.196}$$

That is,

$$[X_a, [X_b, X_c]] + [X_b, [X_c, X_a]] + [X_c, [X_a, X_b]] = 0, \tag{2.197}$$

which when (2.154) is used successively gives:

$$\begin{aligned} i f_{bcd} [X_a, X_d] + i f_{cad} [X_b, X_d] + i f_{abd} [X_c, X_d] = 0, \\ f_{bcd} f_{ade} + f_{cad} f_{bde} + f_{abd} f_{cde} = 0. \end{aligned} \tag{2.198}$$

Also, by the definition of the commutator, $[X, Y] = -[Y, X]$ implies

$$f_{abc} = -f_{bac}. \tag{2.199}$$

If we define a set of matrices, T_a:

$$(T_a)_{bc} \equiv i f_{abc} \tag{2.200}$$

then, (2.198) can be rewritten as:

$$f_{bcd} f_{ade} - f_{acd} f_{bde} = -f_{abd} f_{cde} = f_{abd} f_{dce} \tag{2.201}$$

and using (2.200):

$$-(T_b)_{cd}(T_a)_{de} + (T_a)_{cd}(T_b)_{de} = -i f_{abd} (T_d)_{ce},$$

which implies (due to the matrix multiplication rule):

$$[T_a, T_b]_{ce} = i f_{abd} (T_d)_{ce} \tag{2.202}$$

and finally,

$$[T_a, T_b] = i f_{abd} T_d. \tag{2.203}$$

In other words, the structure constants themselves generate a representation of the algebra. This is called as the *adjoint representation* of the algebra.

2.10 Subgroups, cosets, invariant subgroups

A subgroup H of G is a non-empty subset of G. H itself forms a group with respect to the group composition of G. Thus,

$$e \,\varepsilon\, H, \quad h \,\varepsilon\, G, \quad \Rightarrow h^{-1} \,\varepsilon\, H \quad h_1, h_2 \,\varepsilon\, H. \tag{2.204}$$

Here are a few examples of subgroups of a given group:

- The identity element, e, and the whole group, G, are *trivial* subgroups of G.
- The symmetric group S_3 has three different subgroups consisting of e and permutations of i and j ($i, j = 1,2,3$) alone. That is, the S_2 subgroups of S_3 are:

$$e, \quad P_{12}, \quad P_{13}, \quad P_{23}. \tag{2.205}$$

  P_{ij} are called *transpositions*. All are isomorphic.
- The special orthogonal group of order 3, $SO(3)$, has three distinct $SO(2)$ subgroups. A typical one will be rotations about a fixed axis, $\hat{n}$:

$$\{e^{i(\hat{n}\cdot\vec{J})\theta} \mid \hat{n} \text{ fixed}, \forall\, \theta\}, \tag{2.206}$$

  where $\vec{J}$ is the spin 1 angular momentum operator. The isomorphic map is:

$$e^{i(\hat{n}\cdot\vec{J})\theta} \to e^{i(\hat{m}\cdot\vec{J})\theta}, \quad \hat{n}, \hat{m} \text{ fixed}. \tag{2.207}$$

- Real orthogonal matrices are also unitary. So,

$$O(n) \subset U(n), \qquad SO(n) \subset SU(n). \tag{2.208}$$

  We also have the chain of subsets for any of the above groups, for example:

$$U(n) \supset U(n-1) \supset U(n-2) \supset \cdots \supset U(2) \supset U(1). \tag{2.209}$$

- If G is the Poincaré group, $\mathcal{P}$, the translation group: $T^4 = \{(a, 1)_{a \varepsilon R}\}$ forms an Abelian subgroup of G. The homogeneous Lorentz transformations: $O(3, 1) = \{(0, \Lambda)\}$ form a subgroup of G.

Cosets. Let H be a subgroup of G: $H \subset G$, then the *left coset* of H with respect to the element $g \,\varepsilon\, G$, is the set:

$$g\,H \equiv \{g \cdot h \mid h \,\varepsilon\, H\}. \tag{2.210}$$

Similarly, the right coset is given by:

$$H\,g \equiv \{h \cdot g \mid h \,\varepsilon\, H\}. \tag{2.211}$$

The space of left (right) cosets is $\{g\,H\}_{g \varepsilon G}$ $(\{g\,H\})_{g \varepsilon G}$.

Invariant subgroup. A subgroup $H \subset G$ is an *invariant* or *normal* subgroup if:

$$g\,H\,g^{-1} = H, \qquad \forall\, g \,\varepsilon\, G. \quad (i.e.\ gH = Hg). \tag{2.212}$$

If H is an invariant subgroup of G, the left and right cosets are equal, so, in this case, one can talk of cosets without specifying left or right and denote the space of cosets by:

$$G/H = \{g\, H\} = \{H\, g\}. \tag{2.213}$$

Theorem. If H is an invariant subgroup of G, then G/H is a group called the *factor group* of G with respect to H. The multiplication rule is the set multiplication of cosets. The identity is H. For,

$$g\, H\, g'\, H = g\, g'\, H\, H = g\, g'\, H \text{ (another coset)} \tag{2.214}$$

$g\, H\, H = g\, H$, or H is the identity. Also, $(g\, H)^{-1} = g^{-1}H$ is the inverse since

$$g^{-1}\, H\, g\, H = g^{-1}\, g\, H\, H = H\, H = H. \tag{2.215}$$

Example (i): G the Poincaré group contains $T^4 = \{(a,\, 1)\}$, the four-dimensional translation group as an invarinat abelian subgroup. For,

$$g = \{(0,\, \Lambda)\}\ \varepsilon\ H, \qquad h = \{(a,\, 1)\ \varepsilon\ T^4\} \tag{2.216}$$

$$\begin{aligned} g\, h\, g^{-1} &= (0,\, \Lambda)(a,\, 1)(0,\, \Lambda)^{-1} \\ &= (0,\, \Lambda)(a,\, 1)(0,\, \Lambda^{-1}),\ \text{since } (a,\, \Lambda)^{-1} = (-\Lambda^{-1}a,\, \Lambda^{-1}) \\ &= (0,\, \Lambda)(a,\, \Lambda^{-1}),\ \text{since } (a',\, \Lambda')(a,\, \Lambda) = (\Lambda' a + a',\, \Lambda'\Lambda) \\ &= (\Lambda a + 0,\, \Lambda\Lambda^{-1}) = (\Lambda a,\, 1)\ \varepsilon\ H. \end{aligned} \tag{2.217}$$

Example (ii): the orthogonal group $G = O(3)$ contains the subgroup $H = SO(3)$, which implies that:

$$\det(g\, h\, g^{-1}) = \det h = 1. \tag{2.218}$$

$SO(3)$ is thus an invariant subgroup of $O(3)$.

Example (iii): the Poincaré group $\mathcal{P}$ has the Translation group T^4 as a subgroup. Since

$$T_4(a,\, \Lambda) = T_4(a,\, 1)(0,\, \Lambda) = T_4(0,\, \Lambda) \tag{2.219}$$

and since

$$T_4(0,\, \Lambda)T_4(0,\, \Lambda') = T_4(0,\, \Lambda)(0,\, \Lambda') = T_4(0,\, \Lambda\Lambda') \tag{2.220}$$

the cosets are in one-to-one correspondence with the elements of the Lorentz group $\{\Lambda\}$. G/H is isomorphic to the Lorentz group.

2.11 Simple and semi-simple groups

A group G is *simple* if it has no invariant subgroup besides itself and the identity. Examples of simple groups are S_2 and $SO(3)$.

The unitary unimodular group in two dimensions, $SU(2)$ is *not* simple, since it has an invariant subgroup Z_2:

$$Z_2 = \left\{ \begin{pmatrix} 1 & 0 \\ 0 & 1 \end{pmatrix}; \quad \begin{pmatrix} -1 & 0 \\ 0 & -1 \end{pmatrix} \right\}. \tag{2.221}$$

Similarly, the group $SU(n)$ is not simple, since it has as an invariant subgroup Z_n:

$$Z_n = \left\{ \exp\left(i\frac{2\pi}{n}k \right) \mid k = 0, 1, \ldots, n-1 \right\}. \tag{2.222}$$

A group is *semi-simple* if it has no invariant abelian subgroup besides the identity and itself.

Note. If $g_i g_j = g_j g_i$, for all $g_i, g_j \varepsilon G$, then the group G is abelian. Translations in R^n are abelian. The group $SO(2)$ is abelian.

The Poincaré group is not semi-simple, since the translations form an invariant abelian subgroup.

2.12 Compact groups

Of importance in group theory is the idea of a *compact* group. To understand the concept of compactness, which is a *global* property, we need the concepts of boundedness and closedness.

A set of numbers is said to be *bounded* if none of the numbers in the set exceeds a given positive number M in an absolute value.

A set is said to be *closed* if the limit of every congruent sequence of points in the set also lies in the set.

Examples. The real line between 0 and 1 is bounded. It is open at the end points 0 and 1, that is, 0 and 1 are included in the set. The notation used for the closed set, closed at both ends, is [0,1], while a set that is open at both ends is denoted by (0,1). Sets that are open at one end are called semi-open or semi-closed and are denoted by (0,1] and [0,1).

An r-parameter group is compact if the domain of the variation of all the parameters is bounded and closed.

The group of rotations is compact, since the angle of rotation about any axis must lie in the closed interval $0 \leqslant \theta \leqslant 2\pi$. However, the translation group is not compact, as the parameter a in the transformation $x' = x + a$ can be: $-\infty < a < \infty$. Hence, its domain of variation is unbounded. If the translations a vary over a bounded region, then the translations do not form a group.

The group of Lorentz transformations is characterized by a parameter, the velocity $\vec{v}$, which is bounded by the velocity of light, c. However, the limit point

$v = c$ is not a member of the group as the Lorentz transformations (2.2), which we recall are:

$$x' = \frac{x - vt}{\sqrt{1 - v^2/c^2}}, \qquad t' = \frac{t - vx/c^2}{\sqrt{1 - v^2/c^2}}, \qquad y' = y, \qquad z' = z$$

and they become infinite when $v = c$. Therefore, the Lorentz group is not compact.

A continuous group is *locally compact* if in the neighbourhood of any point in parameter space, a closed domain or interval exists. The point in question can be a boundary point in the domain. A compact group is, of course, locally compact. In addition, some non-compact groups, such as the translation group and the Lorentz group are locally compact.

Cayley's theorem and Lagrange's theorem are two of the fundamental theorems in group theory and they are:

Cayley's theorem[13]: any finite group (of order n) is isomorphic to a subgroup of the group S_n, of permutations of n objects.

Lagrange's theorem: if H is a subgroup of a group G, and if $|G| = n$ (i.e. the order of the finite group G is n), and $|H| = m$, then m is a factor of n.

For a proof of these theorems, we refer the interested reader to Budden's (1972) book *Fascination of Groups*.

An important theorem about group representations is known as:

Schur's Lemma[14]: let G be a finite or compact Lie group. Let D and D' be irreps. of G with dimensions n and n', respectively. Let M be an arbitrary $n' \times n$ matrix (with n' rows and n columns) which is independent of $g \,\varepsilon\, G$. Let

$$M\, D(g) = D'(g)\, M, \qquad \forall \quad g \,\varepsilon\, G. \tag{2.223}$$

Then, if $n' \neq n$, it follows that $M = 0$, while if $n' = n$, either $M = 0$ or M is non-singular and D and D' are equivalent.

Corollary also called Schur's lemma, states that a matrix that commutes with all the matrices of an irrep. is a multiple of the unit matrix.

Note. In this chapter some of the basic concepts and definitions in group theory have been provided as these are essential for an understanding of the contents of the following chapters in this book. The symmetric group has been dealt with at some length, as it is of relevance in the following chapters.

References

Bradley C J and Cracknell A P 1972 *The Mathematical Theory of Symmetry in Solids: Representation Theory for Point Groups and Space Groups* (Oxford: Clarendon)

Budden E 1972 *The Fascination of Groups* (Cambridge: Cambridge Univeristy Press)

Cassidy D C 1991 *Uncertainty: the Life and Times of Werner Heisenberg* (NY: W.H.Freeman)

Crilly T 1995 *The Math. Gazette* **79** 259

---

[13] Arthur Cayley, British mathematician (1821–95). See Crilly (1995).

[14] Named after Issai Schur, Russian mathematician (1875–1941).

Enz C P 2010 *No Time to be Brief: A Scientific Biography of Wolfgang Pauli* (Oxford: Oxford University Press)

Farmelo G 2011 *The Strangest Man: The Hidden Life of Paul Dirac, Mystic of the Atom* (New York: Basic Books)

Federov E S 1971 *Symmetry of Crystals* ed. David and Harker K 7 (Buffalo, NY: American Crystallographic Association), pp 50–131 (transl.)

Galois É 1828 *Ann. Math.* **XIX** 294

Gilmore R 2006 *Lie Groups, Lie Algebras and Some of their Applications* (New York: Dover)

Gray J 2012 *Henri Poincaré A scientific Life* (Princeton, NJ: Princeton University Press)

Greenberg O W 2006 Why is CPT fundamental? *Found. Phys.* **36** 1535–53

Hamermesh M 1962 *Group Theory and its Applications to Physical Problems* (New York: Dover)

Hanc J, Slavomir T and Hancova M 2004 *Am. J. Physics* **72** 428

Hanca J, Tulejab S and Hancova M 2004 *Am. J. Phys* **72** 428–35

Hankins T L 2004 *Sir William Rowan Hamilton* (Baltimore, MD: Johns Hopkins University Press)

Infeld L 1978 *Whom the Gods Love: The Story of Evariste Galois, Classics in Math education* Vol. 6 (Reston, VA: National Council of teachers)

Jacobi C G J 1884 *Gesammelte Werke* ed C W Borchardt, A Clebsch and K Weierstrass (7 vols. and supp.) (Berlin: Prussian Academy of Sciences)

Jackson J D 1975 *Classical Electrodynamics* 2nd edn (New York: Wiley) ch 11, 12,

James W and Smith M K 1981 *Emmy Noether: A Tribute to her Life and Work* (New York: Marcel Dekker)

Kim Y S and Nozi M 2013 *Theory and Applications of the Poincaré Group* (New York: Springer)

Lorentza H A 1904 *Proc. R. Netherlands Aead. Arts Sci.* **6** 809

Lakshminarayanan V and Varadarajan L 2015 *Special Functions for Optical Science and Engineering* (Bellingham, WA: SPIE Press)

Lakshminarayanan V, Ghatak A K and Thyagarajan K 2001 *Lagrangian Optics* (Dordrecht: Kluwer)

Lie S 1869 *Reprasantation der imaginaren der plageometrie* (Christiania, Norway: Acad. Sci.)

Lipkin H J 2002 *Lie Groups for Pedestrians* (New York: Dover)

Mott N and Pierls R 1977 *Bio. Mem. Fellows Roy Soc.* **23** 212

Naimark M A 1964 *Linear Representations of the Lorentz Group* translated by ed A Swinfen and O J Marstrand (London: Pergamon Press)

Noether E 1918 Invariant variation problems *Goot. Nachr.* 235–57. English translation; Tavel M 1971 Invariant variation problems *Transport Theory Stat. Phys.* **1** 186

Ore O 2008 *Neels Henrik Abel* (New York: AMS Chelsea Publishing)

Pepe L 2014 *Lettera math.* **2** 3

Polya G 1974 Collected Papers *Vol 1: Singularities of Analytic Functions; Vol 2: Location of Zeros; Vol 3: Analysis; Vol 4: Probability, Combinatorics* ed R P Boas (Cambridge, MA: MIT Press)

Reid C 1996 *Hilbert* (New York: Springer)

Rigatell L T 1996 *Evariste Galois, Vita Mathematica* (Geneva: Birkhauser)

Rothman T 1982 Genius and Biographers: The fictionalization of Evariste Galvois *Ann. Math. Monthly* **89** 84

Schwichtenberg J 2015 *Physics from Symmetry* (New York: Springer)

Srinivasa Rao K and Pandir P 2016 *J.Ramanujan Soc. math and math sciences* **6** 53

Streater R F and Wightman A S 1964 *PCT, Spin and Statistics and All That* (New York: Benjamin)

Stubhaug A and Daly R 2002 *The Mathematician Sophus Lie* (New York: Springer)

Viana M and Lakshminarayanan V 2013 *Dihedral Fourier Analysis* (Heidelberg: Springer)

Viana M and Lakshminarayanan V 2018 *Symmetry in Optics and Vision Science* (Boca Raton, FL: CRC Press)

Weyl H 1939 *The Classical Groups: Their Invariants and Representations* (Princeton, NJ: Princeton University Press)

Zee A 2016 *Group Theory in a Nutshell for Physicists* (Princeton, NJ: Princeton University Press)

Chapter 3

Group theory of the Kummer solutions of the Gauss differential equation

3.1 Introduction

In chapter 1, the second order differential equation of Gauss was introduced, and the 24 solutions given by Kummer in 1814 were listed in appendix A of that chapter. There are three term recurrence relations amongst these, which are given in Bailey (1935) and Slater (1966).

In chapter 2, we provided an introduction to the concepts and definitions of groups. We are now in a position to establish a connection between the Kummer solutions which can be presented by a single equation. It is a statement of the fact that all the 24 solutions can be obtained from 24 permutations. These correspond to the symmetries of an ordinary cube. The key to this observation is based on relating the four variables of the Gauss hypergeometric function, ${}_2F_1(a,\ b;\ c;\ z)$, to the six parameters, say $f(x_1, x_2, x_3, x_4, x_5, x_6)$, where the x_i, $i = 1, 2, \ldots, 6$ can be associated with the six sides of the cube. These six parameters can then be permuted, and unrestricted, which would result in $6! = 720$ functions $f(x_1, \ldots, x_6)$. Interestingly, 190 years after the 24 solutions were obtained by Kummer in 1814, Lievens, Srinivasa Rao, and Van der Jeugt, in 2005, discovered the single equation from which on superposing the 24 symmetries of the cube, the entire set of Kummer solutions can be obtained. This reveals the power of group theory for unification and simplification.

Bernhard Riemann (1826—1866; see Derbyshire, 2003) is said to have studied the works of Euler and Legendre while still studying in secondary school and also mastered Legendre's treatise on the theory of numbers in less than a week. Gauss' presence in Göttingen made the University of Göttingen a hub for mathematical sciences. However, due to the distance and aloofness of Gauss, Riemann left for the University of Berlin, where he obtained his doctorate in 1851. His dissertation on the general theory of functions of a complex variable—concentrating more on the general principles and geometric ideas—led to him founding his theory on what

doi:10.1088/978-0-7503-1496-1ch3

became renowned as Cauchy–Riemann equations. This theory created the ingenious device of Riemann surfaces for clarifying the nature of multi-valued functions, and also to the Riemann mapping theorem. Even Gauss, reticent in recognizing the significant contributions of his illustrious contemporaries, praised Riemann's work with his statement (Dunnington 2003): 'The dissertation submitted by Herr Riemann offers convincing evidence of the author's thorough and penetrating investigations in those parts of the subject treated in the dissertation, of a creative, active, truly mathematical mind, and of a gloriously fertile originality.' In section 3.3, of this chapter, we present Riemann's generalization of the Gauss second order ordinary differential equation (ODE).

3.2 The 24 solutions of the Gauss ODE

Most textbooks on special functions, such as the ones by Bailey (1935) and Slater (1966), give the second order Gauss ODE:

$$z(1-z)\frac{du^2}{dz^2} + \{c-(a+b+1)\,z\}\frac{du}{dz} - (ab)\,u(z) = 0, \tag{3.1}$$

where a, b, c are real or complex variables, having one solution as the hypergeometric series:

$${}_2F_1(a, b; c; z) = F(a, b; c; z) = \sum_{n=0}^{\infty} \frac{(a)_n\,(b)_n}{(c)n}\frac{z^n}{n!}, \tag{3.2}$$

and $(a)_n$ is the Pochhammer symbol:

$$(a)_n = a(a+1)(a+2)\cdots(a+n-1) = \frac{\Gamma(a+n)}{\Gamma(a)},\ n > 0; \quad (a)_0 = 1, \tag{3.3}$$

listing 24 solutions for (3.1), including (3.1) as the first in that list.

Kummer showed that (3.2) belongs to a set of 24 functions. He published a set of six distinct solutions of the hypergeometric equation, each of which has four forms, related to one another by Euler's transformations:

$$\begin{aligned} F(a, b; c : z) &= (1-z)^{-a}F\left(a, c-b; c; \frac{z}{z-1}\right),\\ F(a, b; c : z) &= (1-z)^{-b}F\left(c-a, b; c; \frac{z}{z-1}\right),\\ F(a, b; c : z) &= (1-z)^{c-a-b}F(c-a, c-b; c; z). \end{aligned} \tag{3.4}$$

Thus, the solution $F(a, b; c; z)$ of the Gauss differential equation (3.1) has four different forms (3.2) and (3.4), or eight different forms if one includes the obvious 'mirror' symmetries:

$$F(a, b; c; z) = F(b, a; c; z), \tag{3.5}$$

giving 24 forms in total. Often these 24 functions are referred to as the Kummer solutions of the Gauss hypergeometric equation.

These 24 Kummer functions (or solutions) are then given as a list and their analytic properties have been extensively discussed in the aforesaid classical textbooks. The fact that these 24 solutions are related to one another by a finite group of transformations was observed by Posser (1994) and in the book by Iwasaki *et al* (1991). However, in none of these references, is this finite group of 24 functions characterized—when the trivial a, b permutations of the numerator variables (3.5) are included, the finite group is 48 functions. As noted previously, the 24 Kummer solutions are related to the finite group of the ordinary cube (see Budden 1972), which is isomorphic to the symmetric group S_4. We shall not refer to any of the analytic properties of the solutions, since these have been discussed extensively in many books, including in Lakshminarayanan and Varadharajan (2015). Kummer's 24 solutions are tabulated in table 3.1 (column 3).

What is being done is to map a function of four variables a, b, c, z onto a function of six variables attached to the six sides of an ordinary cube (such as a die—which incidentally has the numbers 1–6 arranged so that the sum of the opposite sides of the cube (or common die) add to 7: that is, if 1 and 6, 2 and 5 are assigned clockwise to the opposite faces of the dice/cube, then the remaining pair of 3 and 4 can be assigned to either the left or right sides of the cube. Any one of the two choices gives rise to the 24 symmetries. Together we have 48 symmetries). What Leivens *et al* (2005) did was to find the 24 permutations of the symmetric group S_4, given in column 2 of table 3.1.

It has already been established that the 24 Kummer solutions are related to one another by a finite group of the order 24 (or of the order 48, if the mirror symmetries are included; see Iwasaki *et al* 1991 and Posser 1994). It has been shown by Lievens *et al* that the group of 24 is the group of (rotation) symmetries of the cube, also known as the 'direct symmetry group of the cube', or as the octahedral group $\mathcal{O}$ (Miller 1972).

To describe the group $\mathcal{O}$, consider the cube whose faces have been labeled by 1, 2, 3, 4, 5, 6, as stated above (see figure 3.1). Consider the original configuration of the cube, given in figure 3.2(a). Each rotational symmetry of the cube, that is each element of O, is then described by a particular permutation of the faces of the cube. For example, a rotation over $-\pi/2$ among the axis passing through the midpoints of faces 1 and 6 yields the configuration in figure 3.2(b). Clearly, the transformation from figure 3.2(a) to (b) can be described by the permutation:

$$\begin{pmatrix} 1 & 2 & 3 & 4 & 5 & 6 \\ 1 & 4 & 2 & 5 & 3 & 6 \end{pmatrix}, \tag{3.6}$$

or in the equivalent cycle notation (2,4,5,3). Similarly, figure 3.2(c) describes a rotation through the axis passing through the midpoint of faces 3 and 4, the corresponding permutation of the faces of the cube is given by (1, 2, 6, 5). The group $\mathcal{O}$ is thus a subgroup of the symmetric group S_6. Its 24 elements are given in the second column of table 3.1. It is now easy to see that $\mathcal{O}$ is the subgroup of S_6 generated by the two elements:

$$g_1 = (2, 4, 5, 3) \qquad \text{and} \qquad g_2 = (1, 2, 6, 5). \tag{3.7}$$

Table 3.1. Kummer's 24 solutions.

i	Permutation	Function	α_i
1	()	$F(a, b; c; z)$	0
	(2,4,5,3)	$(1-z)^{-b}F\left(b, c-a; c; \frac{z}{z-1}\right)$	
	(2,5)(3,4)	$(1-z)^{c-a-b}F(c-a, c-b; c; z)$	
	(2,3,5,4)	$(1-z)^{-a}F\left(c-a, a; c; \frac{z}{z-1}\right)$	
2	(1,2,6,5)	$z^{1-c}F(1+a-c, 1+b-c; 1+a+b-c; 1-z)$	$\frac{1}{4}(a+b)-\frac{1}{12}$
	(1,2)(3,4)(5,6)	$F(a, b; 1+a+b-c; 1-z)$	
	(1,2,3)(4,6,5)	$z^{-a}F(a, c-b; 1+a-b; 1+a+b-c; 1-\frac{1}{z})$	
	(1,2,4)(3,6,5)	$z^{-b}F(1+b-c, b; 1+a+b-c; 1-\frac{1}{z})$	
3	(1,3,6,4)	$z^{-a}F\left(a, 1+a-c; 1+a-b; \frac{1}{z}\right)$	$\frac{1}{4}(3a+b-c)+\frac{1}{12}$
	(1,3)(2,5)(4,6)	$z^{b-c}(1-z)^{c-a-b}F\left(c-b, 1-b; 1+a-b; \frac{1}{z}\right)$	
	(1,3,2)(4,5,6)	$(1-z)^{-a}F\left(a, c-b; 1+a-b; \frac{1}{1-z}\right)$	
	(1,2,5)(2,6,4)	$z^{1-c}(1-z)^{c-a-1}F\left(1-b, 1+a-c; 1+a-b; \frac{1}{1-z}\right)$	
4	(1,4,6,3)	$z^{-b}F\left(b, 1+b-c; 1+b-a; \frac{1}{z}\right)$	$\frac{1}{4}(a+3b-c)+\frac{1}{12}$
	(1,4)(2,5)(3,6)	$z^{a-c}(1-z)^{c-a-b}F\left(1-a, c-a; 1+b-a; \frac{1}{z}\right)$	
	(1,4,2)(3,5,6)	$(1-z)^{-b}F\left(c-a, b; 1-a+b; \frac{1}{1-z}\right)$	
	(1,4,5)(2,6,3)	$z^{1-c}(1-z)^{c-b-1}F\left(1-a, 1+b-c; 1-a+b; \frac{1}{1-z}\right)$	
5	(1,5,6,2)	$(1-z)^{c-a-b}F(c-a, c-b; 1-a-b+c; 1-z)$	$-\frac{1}{4}(a+b-2c)-\frac{1}{12}$
	(1,5)(2,6)(3,4)	$z^{1-c}(1-z)^{c-a-b}F(1-a, 1-b; 1-a-b+c; 1-z)$	
	(1,5,3)(2,4,6)	$z^{a-c}(1-z)^{c-a-b}F\left(1-a, c-a; 1-a-b+c; 1-\frac{1}{z}\right)$	
	(1,5,4)(2,3,6)	$z^{b-c}(1-z)^{c-a-b}F(1-b, c-b; 1-a-b+c; \frac{1}{1-z})$	
6	(1,6)(2,5)	$z^{1-c}(1-z)^{c-a-b}F(1-a, 1-b; 2-c; z)$	$\frac{c}{2}-\frac{1}{2}$
	(1,6)(3,4)	$z^{1-c}F(1+a-c, 1+b-c; 2-c; z)$	
	(1,6)(2,4)(3,5)	$z^{1-c}(1-z)^{c-b-1}F\left(1-a, 1+b-c; 2-c; \frac{z}{z-1}\right)$	
	(1,6)(2,3)(4,5)	$z^{1-c}(1-z)^{c-a-1}F\left(1+a-c, 1-b; 2-c; \frac{z}{z-1}\right)$	

Also, observe that $\mathcal{O}$ is isomorphic to the symmetric group S_4. This can be seen by considering the four main diagonals of the cube: every symmetry is then uniquely described by a permutation of these four diagonals (see Budden 1972).

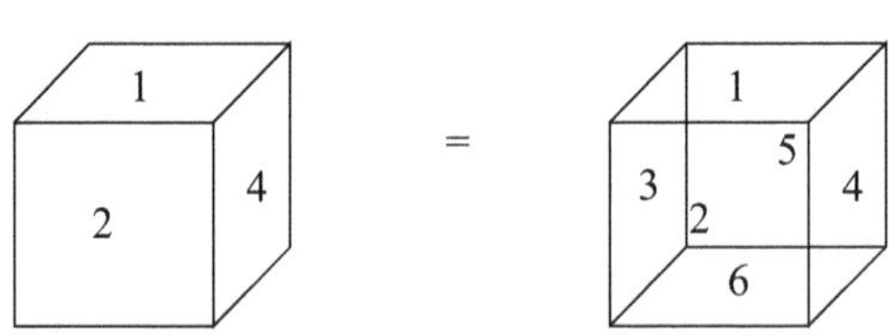

Figure 3.1. The cube with the top face 1 and bottom face 6, front face 2 and back face 5, left face 3 and right face 4.

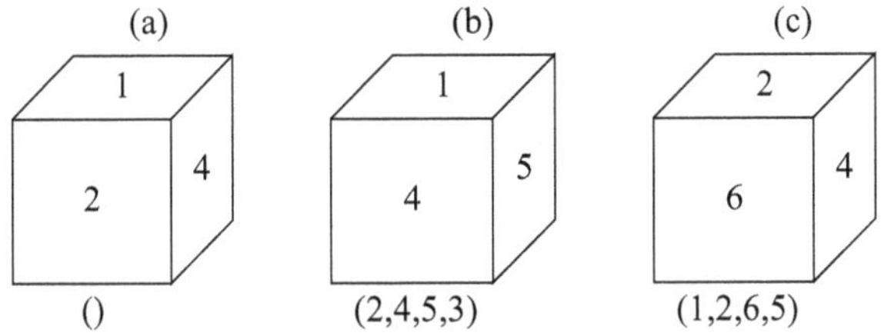

Figure 3.2. The cube and two of the rotation symmetries.

Let x_i, $(i = 1, 2, \ldots, 6)$ be six variables associated with the six sides of a cube, satisfying the constraint:

$$\sum_{i=1}^{6} x_i = 0. \tag{3.8}$$

Consider the function of six variables:

$$f(x) = F\left(\frac{1}{2} + x_1 + x_2 + x_3, \frac{1}{2} + x_1 + x_2 + x_4; \; 1 + x_1 - x_2; -\frac{x_1 + x_6}{x_3 + x_4}\right). \tag{3.9}$$

Let us identify the four variables of the Gauss hypergeometric function ${}_2F_1(a, b; c; z)$ with a, b, c, z. Solve this system with respect to the six variables x_i. Because of the constraint (3.8) on the six variables, only five of the six x_i are independent. This leaves us now with one degree of freedom. Consider any element of g of the group $\mathcal{O}$. The action of g on the function x is determined by permuting the indices of the x_i. So, acting with the element defined above $g_1 = (2, 4, 5, 3)$ on $f(x)$ gives:

$$f(g_1 \cdot x) = {}_2F_1\left(\frac{1}{2} + x_1 + x_2 + x_4, \frac{1}{2} + x_1 + x_4 + x_5; \; 1 + x_1 - x_6; -\frac{x_1 + x_6}{x_2 + x_5}\right). \tag{3.10}$$

This is equal to ${}_2F_1\left(b, c - a; c; \frac{z}{z-1}\right)$ when the original $f(x)$ is identified with ${}_2F_1(a, b; c; z)$.

Similarly, we find that with

$$\begin{aligned} g_2 &= (1, 2, 6, 5), \\ f(g_2 \cdot x) &= {}_2F_1(a + a - c, 1 + b - c; a + a + b - c; 1 - z). \end{aligned} \tag{3.11}$$

For each element g of $\mathcal{O}$, the corresponding function $f(g{\cdot}x)$ is given in the third column of table 3.1—(up to the variables in the hypergeometric function and not yet taking the powers of z and $1 - z$ into account).

Thus, with every element of the group $\mathcal{O}$, one of the 24 Kummer solutions is associated. We can do even better and reproduce the complete solution by including the power functions multiplying the hypergeometric functions to technically complete the correspondence between the symmetries of the cube and the 24 solutions of Kummer. The symbolic system, Mathematica (Wolfram Research, Champaign, IL), was used to arrive at the function that fills the bill:

$$C(x) = (-1)^{\frac{x_1 - x_6}{4} + \frac{1}{12}}(x_1 + x_6)^{\frac{x_1 - x_6}{2} + \frac{1}{6}}(x_2 + x_5)^{\frac{x_2 - x_5}{2} + \frac{1}{6}} \times (-x_3 - x_4)^{\frac{x_1 - x_6}{2} + \frac{x_2 - x_5}{2} + \frac{1}{3}}. \tag{3.12}$$

The action of an element $g \varepsilon \mathcal{O}$ is again by permutation of the indices:

$$g : f(x) \to \frac{C(g \cdot x)}{C(x)} f(g \cdot x). \tag{3.13}$$

It is now easy to verify that the 24 elements of the group $\mathcal{O}$ yield the 24 solutions of Kummer given in chapter 1.

For completeness (3.13) gives a required additional constant multiplying the Kummer solution. This constant factor is $(-1)^{\alpha}{}_{i}$, and it is this factor that finds a place in column 4 of table 3.1. Note that it is the same for four solutions given in the six sets (of four solutions each).

The mirror symmetries, also sometimes referred to in literature as Yutsis symmetries (as in Smorodinskii and Shelepin 1972), are not included table 3.1. If we extend the group $\mathcal{O}$ by including the reflection symmetries of the cube, it now becomes a group of the order 48, and we refer to it as $\mathcal{O}_{\langle}$. This extension can be done by adding the additional generator $g_3 = (3, 4)$ to the two generators, g_1 and g_2. In figure 3.1, we can see that g_3 corresponds to a reflection about a plane. In (3.9), the permutation of the indices (3,4) corresponds to the mirror symmetry (3.5). One can verify that the 48 elements of the extended group $\mathcal{O}_{\langle}$ yields the 24 Kummer solutions, and their mirror symmetries.

The subgroup of $\mathcal{O}$, consisting of those symmetries that leave the top and the bottom face (with the labels 1 and 6) invariant, is a cyclic group of order 4. This group describes the four solutions related to one another by the transformations of Euler, (3.2) and (3.4).

3.3 The Riemann equation

The Gauss second order ODE is related to the Riemann equation:

$$\begin{aligned} u''(z) &+ \left(\frac{1 - \alpha - \alpha'}{z - z_a} + \frac{1 - \beta - \beta'}{z - z_b} + \frac{1 - \gamma - \gamma'}{z - z_c} \right) u'(z) \\ &+ \left(\frac{\alpha\alpha'(z_a - z_b)(za - z_c)}{z - z_a} + \frac{\beta\beta'(z_b - z_c)(z_b - z_a)}{z - z_b} \right. \\ &\left. + \frac{\gamma\gamma'(z_c - z_a)(z_c - z_b)}{z - z_c} \right) \\ &\times \frac{u(z)}{(z - z_z)(z - z_b)(z - z_c)} = 0, \end{aligned} \tag{3.14}$$

where

$$\alpha + \alpha' + \beta\beta' + \gamma\gamma' = 1. \tag{3.15}$$

Putting the regular singularities

$$(z_a, z_b, z_c) = (0, 1, \infty) \tag{3.16}$$

and

$$(\alpha, \alpha'), (\beta, \beta'), (\gamma, \gamma') \text{ to } (0, 1 - c), (a, b), (0, c - a - b), \tag{3.17}$$

we obtain the Gauss second order ODE (3.1).

A solution of the Riemann equation (3.14) is given by:

$$\begin{aligned}&\left(\frac{z - z_a}{z - z_b}\right)^{\alpha}\left(\frac{z - z_c}{z - z_b}\right)^{\gamma} \times {}_2F_1\Bigg(\alpha + \beta + \gamma, \alpha + \beta' + \gamma; \alpha + \beta' + \gamma;\\ &\qquad 1 + \alpha - \alpha'; \frac{(z_c - z_b)(z - z_a)}{(z_c - z_a)(z - z_b)}\Bigg).\end{aligned} \tag{3.18}$$

In total, we can list 24 such solutions, as in Whittaker and Watson (1965), or as in the case of the Gauss second order ODE (3.1), 48 solutions when the mirror symmetries are included. These 48 solutions arise from the $3! = 6$ permutations of the singularities (z_a, z_b, z_c) and the $2^3 = 8$ transpositions of the primed and unprimed parameters (α, α'), (β, β'), (γ, γ') in the following way. Consider again the cube, with opposite faces now being labeled by α and α', β and β', γ and γ'. The axis passing through the midpoints of the faces α and α' is labeled z_a, β; β' is labeled z_b, and the faces γ, γ' labeled by z_c.

The 48 solutions arise from the $3! = 6$ permutations of the singularities (z_a, z_b, z_c) and the $2^3 = 8$ transpositions of the primed and unprimed parameters (α, α'), (β, β'), (γ, γ'), in the following way. Consider again the cube, with opposite faces being now labeled α and α', β and β', γ and γ'. The axis passing through the midpoints of the faces α and α' is labelled z_α and similarly z_β and z_γ, as in figure 3.3 below.

The 24 (or, including the mirror symmetries, 48) solutions of the Riemann equation (3.14) are in an obvious way related to the symmetries of the cube. Here, this follows immediately from the fact that each of the symmetries of the cube in figure 3.3 leaves the equation (3.14) invariant.

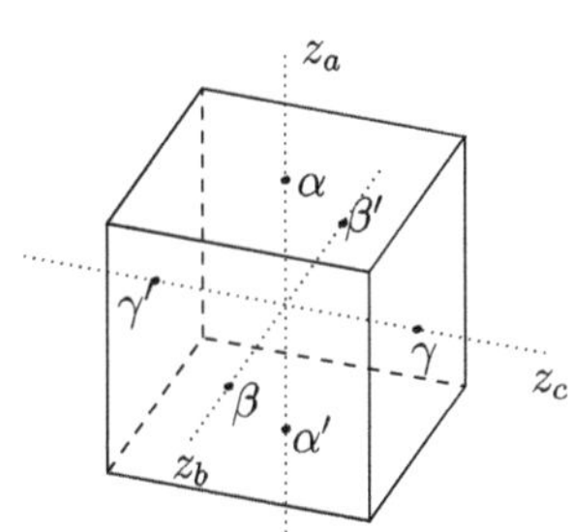

Figure 3.3. The cube with labels of faces and axes referring to the Riemann equation.

References

Bailey W N 1935 *Generalized Hypergeometric Series*Cambridge Tracts in Mathematics and Mathematical Physics 32 (Cambridge: Cambridge University Press)

Budden 1972 *The Fascination of Groups* (Cambridge: Cambridge University Press)

Derbyshire J 2003 *Prime Obsession: Bernhard Riemann and the Greatest Unsolved Problem in Mathematics* (Washington DC: Joseph henry Press)

Dunnington G W 2003 Carl Friedrich Gauss: Titan of Science: A Study of his Life and Work (New York: Hafner)

Iwasaki K, Kimura H, Shimomura S and Yoshida M 1991 From Gauss to Painleve. A Modern Theory of Special Functions *Aspects of Mathematics* (Wiesbaden: Vieweg)

Lakshminarayanan V and Varadarajan L 2015 *Special Functions for Optical Science and Engineering* (Washington, DC: SPIE)

Lievens S, Rao K S and Vander Jeugt J 2005 *Integral Transform Spec. Funct.* **16** 153–8

Miller W Jr 1972 *Symmetry Groups and their Applications* (New York: Academic)

Posser R T 1994 *Am. Math. Monthly* **101** 535–43

Slater L J 1966 *Generalized Hypergeometric Functions* (Cambridge: Cambridge University Press)

Smorodinskii Ya A and Shelepin Leonid A 1972 *Sov. Phys. Uspekhi* **15** 1–24

Whittaker E T and Watson G N 1965 *A Course of Modern Analysis* (Cambridge: Cambridge University Press)

Chapter 4

Group theory of terminating and non-terminating ${}_3F_2(a, b, c; d, e; 1)$ transformations

4.1 Introduction

The hypergeometric function ${}_2F_1(a, b; c; z)$ is defined by the series

$$ {}_2F_1(a, b; c; z) = \sum_{m=0}^{\infty} \frac{(a)_m (b)_m}{(c)_m} \frac{z^m}{m!} \tag{4.1} $$

and, as pointed out by Gauss, a function of four variables a, b, c, z. When one of the variables in the numerator is a negative integer, say $-n$, the series terminates, since the Pochammer symbol

$$ (-n)_m = (-n)(-n+1)(-n+2)\cdots 2\cdot 1\cdot 0 = 0, \quad \forall n, m \,\varepsilon\, \mathcal{Z}. \tag{4.2} $$

Though the generalized hypergeometric function, ${}_pF_q(z)$, $p, q \,\varepsilon\, \mathcal{Z}$, can have an arbitrary number of numerator and denominator variables and is denoted by

$$ {}_pF_q(a_1, a_2, \ldots, a_p; b_1, b_2, \ldots b_q; z) \equiv {}_pF_q\left(\begin{matrix} a_1, a_2, \ldots a_p; z \\ b_1, b_2, \ldots b_q \end{matrix}\right), \tag{4.3} $$

in quantum theory of angular momentum (QTAM; Srinivasa Rao and Rajeswari 1993), those of physical significance are the ones with one numerator variable more than the denominator variables, the same applies too with $z = 1$—referred to as the hypergeometric function with a unit argument, denoted by

$$ {}_{p+1}F_p(a_1, a_2, \ldots, a_p, a_{p+1}; b_1, b_2, \ldots, b_p; 1) = {}_{p+1}F_p(1). \tag{4.4} $$

In QTAM, the ${}_3F_2(1)$ is related to a coefficient that arises in the coupling of two angular momenta, $\vec{j_1}, \vec{j_2}$. The ${}_4F_3(1)$ is related to a coefficient, which arises as an orthogonal transformation between any two of three different ways of coupling the

doi:10.1088/978-0-7503-1496-1ch4

three angular momenta $\vec{j_1}, \vec{j_2}, \vec{j_3}$; coupled two at a time with an intermediate angular momentum $\vec{j_{12}}$ or $\vec{j_{23}}$, for example, in two of the three ways. These coefficients are essential in all the evaluations of matrix elements for spherical tensor operators, denoted in the bra-ket notation of Dirac as:

$$\langle j_f \; m_f \; | Y^{\ell}_{ml} | \; j_i \; m_i \rangle, \tag{4.5}$$

where the Y^{ℓ}_{m} is a spherical tensor of rank ℓ, its projection quantum number $-\ell \leqslant m \leqslant \ell$, and the subscripts i and f refer to the initial and final angular momentum states, which are orthonormal (see for example, Smorodinskii and Shelepin 1972, Biedenharn and Louck 1984).

Beyer, Louck and Stein (1987) showed that an identity (Thomae (1879)) between two ${}_3F_2(1)$ series'—together with invariance under separate permutations of the numerator and denominator variables—implies that the symmetric group S_5 is an invariance group of the non-terminating series. In the same paper, Bailey's transformation for the terminating Saalshültzian ${}_4F_3$ series (Bailey 1935, p 56) is used to study the symmetry group of two-term relations for the series, which is also the symmetric group S_5.

Weber and Erdelyi (1952), showed that a transformation for a terminating ${}_3F_2(1)$ series, when used on itself, again gives rise to a second transformation. The question of continuing the recursive use of the transformation led Srinivasa Rao *et al* (1992) to generate a set of 18 transformations of the terminating ${}_3F_2(1)$ series. Since the set of transformations of the hypergeometric series is invariant to the trivial permutations of the numerator and denominator variables, in the case of the ${}_3F_2(1)$ we are led to a set of $18 \times 2 \times 2 = 72$ transformations, which have been shown to give rise to a *new* 72-element group of transformations, denoted by G_T. The group is new since the group is not isomorphic to a subgroup of S_n of permutations of n objects, as is required by Cayley's theorem, for any finite group.

This 72-element finite group of transformations has nine conjugacy classes and, correspondingly, nine irreducible representations (irreps): four of dimension one, one of dimension two, and four of dimension four. The three generators for this 72-element group are given. The smallest invariant or normal subgroup, H_9 (say), of this finite group is of order 9, is imbedded in an 18 element invariant subgroup. H_{18} and H_{18}, in turn, are imbedded in three 36 element invariant subgroups. In terms of Whipple's parameters (1925) for the ${}_3F_2(1)$, it is shown that H9 is isomorphic to the product of two cyclic groups of order 3.

The 72-element group G_T is shown to be of the invariance group ${}_3F_2$, which is a rescaling of the terminating ${}_3F_2(1)$. Thus, it is the group generating all the two-term relations for this series. The phase factor appearing in such a two-term relation is shown to be equal to an irreducible character of G_T, motivating the construction of the complete character table for G_T. In section 4.2, the essential notation required—due to Whipple—is given. In section 4.3, the 18 terminating ${}_3F_2(a, b, c; d, e; 1)$ transformations are given. Starting with a matrix representing the Weber–Erdelyi transformation for a terminating ${}_3F_2(1)$ series, the procedure for generating the 72-element group G_T is presented. In section 4.4, the structure of the group G_T—its conjugacy classes, its irreps, and their corresponding characters—and the invariant subgroups of G_T are presented. In section 4.5, comments and conclusions regarding a scaling transformation that makes G_T an invariance group of the terminating ${}_3F_2(1)$ series are made; also for the use of the symmetry group in the context of the 3-j coefficient. In section 4.6, the 18 transformations of the ${}_3F_2(1)$ are stated in the Whipple notation; and they are done so in a scaled and invariant form.

4.2 The Whipple notation

Whipple (1925) introduced six parameters $_ri$, $i = 0, 1, 2, 3, 4, 5$, such that:

$$\sum_{i=0}^{5} r_i = 0 \tag{4.6}$$

and let

$$\alpha_{\ell mn} = \frac{1}{2} + r_\ell + r_m + r_n, \qquad \beta_{mn} = 1 + r_m - r_n. \tag{4.7}$$

With these he defined the function:

$$F_p(\ell; mn) = \frac{1}{\Gamma(\alpha_{ijk}, \beta_{m\ell}, \beta_{n\ell})}\, {}_3F_2(\alpha_{ijk}, \alpha_{jmn}, \alpha_{kmn}; \beta_{m\ell}, \beta_{n\ell}; 1) \tag{4.8}$$

where i, j and k are used to represent those three numbers out of the six integers 0, 1, 2, 3, 4, 5 not already represented by ℓ, m, n. The function ${}_3F_2(1)$ is the generalized hypergeometric function (Slater 1966) of unit argument having $\alpha_{imn}, \alpha_{jmn}, \alpha_{kmn}$ as its three numerator variables and $\beta_{m\ell}, \beta_{n\ell}$ as its two denominator variables. By changing the signs of all the r_i parameters and using the constraint (4.6), Whipple defined another function:

$$F_n(\ell; mn) = \frac{1}{\Gamma(\alpha_{\ell mn}, \beta_{\ell m}, \beta_{\ell n})}\, {}_3F_2(\alpha_{\ell jk}, \alpha_{\ell ik}, \alpha_{\ell ij}; \beta_{\ell m}, \beta_{\ell n}; 1). \tag{4.9}$$

In (4.8) and (4.9) use is made of the notation:

$$\Gamma(x, y, z, \ldots) = \Gamma(x)\ \Gamma(y)\, \Gamma(z)\cdots \tag{4.10}$$

By permutation of the suffixes ℓ, m, n over the six integers 0, 1, 2, 3, 4, 5, 60 F_p functions and 60 F_n functions can be written down. If there is no negative integer in the numerator parameters, these series converge only if the real parts of i, j, k in (4.8) and ℓ, m, n in (4.9) are positive. For the sake of brevity the unit argument of the generalized hypergeometric series will not be displayed and it will be denoted as ${}_3F_2(a, b, c; d, e)$, with three numerator and two denominator variables[1].

[1] The use of n as a suffix for the F_n function and also as an index for α and β is continued as in the literature.

4.3 Terminating ${}_3F_2(1)$ series

Consider the transformation for a terminating ${}_3F_2$ used by Weber and Erdelyi (1952):

$$ {}_3F_2\left(\begin{matrix} a,\ b,\ -N \\ d,\ e \end{matrix}\right) = \frac{\Gamma(d,\ d+N-a)}{\Gamma(d+N,\ d-a)}\,{}_3F_2\left(\begin{matrix} a,\ e-b,\ -N \\ 1+a-d-N,\ e \end{matrix}\right). \tag{4.11} $$

The proof for this formula, one of a set (Bailey 1935) obtained by Whipple (1925), is based on the beta integral representation for the ${}_3F_2(1)$:

$$ {}_3F_2(a,\ b,\ -n;\ d,\ e;\ 1) = \frac{\Gamma(c)}{\Gamma(b,\ d-c)}\int_0^1 {}_2F_1(a,\ -n;\ c;\ t)t^{b-1}(1-t)^{d-b-a}dt, \tag{4.12} $$

which can be verified by expanding both sides in the power series. Now for the ${}_2F_1$ we will use the well-known identity:

$$ {}_2F_1(a,\ -b;\ c;\ 1) = \frac{\Gamma(c,\ c+n-a)}{\Gamma(c+n,\ c-a)}\,{}_2F_1(a,\ -n;\ a-n-c+1;\ 1-t). \tag{4.13} $$

Substituting (4.13) in (4.12), replacing t with $1-t$, and using (4.12) again to replace the integral with the ${}_3F_2(1)$, we get the Weber–Erdelyi transformation (4.11) for a ${}_3F_2(1)$.

The formula (4.11) is one of a set (Bailey 1935) obtained by Whipple (1925). If the five parameters of the ${}_3F_2$ on the left-hand side of (4.11) are denoted by the column vector:

$$ \vec{x} = (a,\ b,\ 1-N,\ d,\ e), \tag{4.14} $$

then the parameters of the ${}_3F_2$ on the right-hand side of (4.11) are obtained when the matrix

$$ g_1 = \begin{bmatrix} 1 & 0 & 0 & 0 & 0 \\ 0 & -1 & 0 & 0 & 1 \\ 0 & 0 & 1 & 0 & 0 \\ 1 & 0 & 1 & -1 & 0 \\ 0 & 0 & 0 & 0 & 1 \end{bmatrix} \tag{4.15} $$

operates on $\vec{x}$. Note that $1-N$ is used instead of $-N$, as a component of the column vector $\vec{x}$, since it represents the number of terms in a terminating series. However, ${}_3F_2(a,\ b,\ -N;\ d,\ e)$ will be denoted by ${}_3F_2(\vec{x})$.

Using (4.11) again, with the roles of d and e interchanged, to transform the right-hand side of (4.11), Weber and Erdelyi obtained the transformation:

$$ \begin{aligned} {}_3F_2\left(\begin{matrix} a,\ b,\ -N \\ d,\ e \end{matrix}\right) &= \frac{\Gamma(d,\ e,\ e+N-a,\ d+N-a)}{\Gamma(d+N,\ e+N,\ d-a,\ e-a)} \\ &\quad \times {}_3F_2\left(\begin{matrix} a,\ 1-s,\ -N \\ 1-b+d-s,\ 1-b+e-s \end{matrix}\right), \end{aligned} \tag{4.16} $$

where $s = d + e - a - b + N$. The question arises as to whether this recursive use of the Weber–Erdelyi transformation (4.11) can be continued. In fact, such a procedure, when continued *ad nauseum,* results in a group of 72 transformations—made up of the following 18 terminating ${}_3F_2$ transformations on which the trivial S_2 transformations of

$$a \leftrightarrow b, \qquad \text{and} \qquad d \leftrightarrow e \tag{4.17}$$

are superposed.

$${}_3F_2\begin{pmatrix} a,\, b,\, -N \\ d,\, e \end{pmatrix} = \frac{(d-a, N)}{(d, N)}\, {}_3F_2\begin{pmatrix} a,\, e-b,\, -N \\ 1+a-d-N,\, e \end{pmatrix} \tag{4.18}$$

$$= (-1)^N \frac{(1-s, N)}{(d, N)}\, {}_3F_2\begin{pmatrix} e-a,\, e-b,\, -N \\ s-N,\, e \end{pmatrix} \tag{4.19}$$

$$= \frac{(d-a, N)(e-a, N)}{(d, N)(e, N)}\, {}_3F_2\begin{pmatrix} a,\, 1-s,\, -N \\ 1+a-d-N,\, 1+a-e-N \end{pmatrix} \tag{4.20}$$

$$= \frac{(d-a, N)(b, N)}{(d, N)(e, N)}\, {}_3F_2\begin{pmatrix} e-b,\, 1-d-N,\, -N \\ 1-b-N,\, 1+a-d-N \end{pmatrix} \tag{4.21}$$

$$= \frac{(d-b, N)}{(d, N)}\, {}_3F_2\begin{pmatrix} e-a,\, b,\, -N \\ 1+b-d-N,\, e \end{pmatrix} \tag{4.22}$$

$$= (-1)^N \frac{(1-s, N)(b, N)}{(d, N)(e, N)}\, {}_3F_2\begin{pmatrix} e-b,\, d-b,\, -N \\ 1-b-N,\, s-N \end{pmatrix} \tag{4.23}$$

$$= (-1)^N \frac{(1-s, N)(a, N)}{(d, N)(e, N)}\, {}_3F_2\begin{pmatrix} e-a,\, d-a,\, -N \\ 1-a-N,\, s-N \end{pmatrix} \tag{4.24}$$

$$= (-1)^N \frac{(d-a, N)(d-b, N)}{(d, N)(e, N)}\, {}_3F_2\begin{pmatrix} 1-s,\, 1-d-N,\, -N \\ 1+a-d-N,\, 1+b-d-N \end{pmatrix} \tag{4.25}$$

$$= (-1)^N \frac{(e-a, N)}{(e, N)}\, {}_3F_2\left(\begin{matrix} a, d-b, -N \\ d, 1+a-e-N \end{matrix}\right) \tag{4.26}$$

$$= (-1)^N \frac{(e-a, N)(e-b, N)}{(d, N)(e, N)}\, {}_3F_2\left(\begin{matrix} 1-s, 1-e-N, -N \\ 1+a-e-N, 1+b-e-N \end{matrix}\right) \tag{4.27}$$

$$= (-1)^N \frac{(a, N)(b, N)}{(d, N)(e, N)}\, {}_3F_2\left(\begin{matrix} 1-d-N, 1-e-N, -N \\ 1-a-N, 1-b-N \end{matrix}\right) \tag{4.28}$$

$$= {}_3F_2\left(\begin{matrix} a, b, -N \\ d, e \end{matrix}\right) \qquad \text{(identity)} \tag{4.29}$$

$$= \frac{(d-b, N)(a, N)}{(d, N)(e, N)}\, {}_3F_2\left(\begin{matrix} e-a, 1-d-N, -N \\ 1-a-N, 1+b-d-N \end{matrix}\right) \tag{4.30}$$

$$= \frac{(d-b, N)(e-b, N)}{(d, N)(e, N)}\, {}_3F_2\left(\begin{matrix} b, 1-s, -N \\ 1+b-d-N, 1+b-e-N \end{matrix}\right) \tag{4.31}$$

$$= \frac{(1-s, N)}{(e, N)}\, {}_3F_2\left(\begin{matrix} d-a, d-b, -N \\ d, s-N \end{matrix}\right) \tag{4.32}$$

$$= \frac{(b, N)(e-a, N)}{(d, N)(e, N)}\, {}_3F_2\left(\begin{matrix} d-b, 1-e-N, -N \\ 1+a-e-N, 1-b-N \end{matrix}\right) \tag{4.33}$$

$$= \frac{(a, N)(e-b, N)}{(d, N)(e, N)}\, {}_3F_2\left(\begin{matrix} d-a, 1-e-N, -N \\ 1-a-N, 1+b-e-N \end{matrix}\right) \tag{4.34}$$

$$= \frac{(e-b, N)}{(e, N)}\, {}_3F_2\left(\begin{matrix} b, d-a, -N \\ d, 1+b-e-N \end{matrix}\right). \tag{4.35}$$

These transformations reduce to five relations when they are written in terms of Whipple parameters and the notation of Whipple given in section 4.2. They are:

$$\Gamma(\alpha_{123}, \alpha_{124}, \alpha_{125})F_p(0) = \Gamma(\alpha_{023}, \alpha_{024}, \alpha_{025})F_p(1) \tag{4.36}$$

$$= \Gamma(\alpha_{013}, \alpha_{014}, \alpha_{015})F_p(2) \tag{4.37}$$

$$= (-1)^N\Gamma(\alpha_{123}, \alpha_{013}, \alpha_{023})F_n(3) \tag{4.38}$$

$$= (-1)^N\Gamma(\alpha_{124}, \alpha_{014}, \alpha_{024})F_n(4) \tag{4.39}$$

$$= (-1)^N\Gamma(\alpha_{125}, \alpha_{015}, \alpha_{025})F_n(5) \tag{4.40}$$

where (4.36) represents (4.30), (4.31) and (4.34), (4.37) represents (4.20), (4.21) and (4.33), (4.38) represents (4.23), (4.24) and (4.28), (4.39) represents (4.26), (4.27),and (4.35), (4.40) represents (4.18), (4.22), and (4.25); while (4.29) is the identity; (4.19) and (4.33) correspond to

$$F_p(0; 45) = F_p(0; 35) \quad \text{and} \quad F_p(0; 45) = F_p(0; 34), \tag{4.41}$$

respectively. These relations:

$$F_p(0; 45) = F_p(0; 35) = F_p(0; 34) \tag{4.42}$$

represent the fact that for a given l, all ten expressions $F_p(l; mn)$ (as well as, all the ten $F_n(l; mn)$) are equal. It is for this reason that they are denoted simply as $F_p(l)$ or $F_n(l)$ above. The relations (4.36)–(4.40) are the same as (4.3.3.2)–(4.3.3.6) in Slater (1966), who has also tabulated the expressions for α (and β) in terms of a, b, c (= $-N$), d, e (cf table 4.1, Slater 1966). The transformation (4.28) represents the reversal of the series.

If the scaling transformation (29) is used in the definitions (4.8) and (4.9) for the $F_p(l; mn)$ and $F_n(l; mn)$ functions, then for $\alpha_{kmn} = -N$:

$$F_p(l; mn) = \frac{1}{\Gamma(\alpha_{ijk}, \alpha_{ijm}, \alpha_{ijn})}\, {}_3\tilde{F}_2\begin{pmatrix} \alpha_{imn}, \alpha_{jmn}, -N \\ \beta_{ml}, \beta_{nl} \end{pmatrix} \tag{4.43}$$

and

$$F_n(l; mn) = \frac{1}{\Gamma(\alpha_{lmn}, \alpha_{kln}, \alpha_{klm})}\, {}_3\tilde{F}_2\begin{pmatrix} \alpha_{ljk}, \alpha_{lik}, -N \\ \beta_{lm}, \beta_{ln} \end{pmatrix}. \tag{4.44}$$

Redefining:

$$\tilde{F}_p(l; mn) = \Gamma(\alpha_{ijk}, \alpha_{ijm}, \alpha_{ijn})F_p(l; mn), \tag{4.45}$$

and

$$\tilde{F}_n(l;\, mn) = \Gamma(\alpha_{lmn},\, \alpha_{kln},\, \alpha_{klm})F_n(l;\, mn), \tag{4.46}$$

for $\alpha_{345} = -N$, the relations (4.18) to (4.22) will now become simply:

$$\tilde{F}_p(0) = \tilde{F}_p(1) = \tilde{F}_p(2) \tag{4.47}$$

$$= (-1)^N \tilde{F}_n(3) = (-1)^N \tilde{F}_n(4) = (-1)^N \tilde{F}_n(5), \tag{4.48}$$

since

$$\tilde{F}_p(0) = \Gamma(\alpha_{123},\, \alpha_{124},\, \alpha_{125})F_p(0), \tag{4.49}$$

$$\tilde{F}_p(1) = \Gamma(\alpha_{023},\, \alpha_{024},\, \alpha_{025})F_p(1), \tag{4.50}$$

$$\tilde{F}_p(2) = \Gamma(\alpha_{013},\, \alpha_{014},\, \alpha_{015})F_p(2), \tag{4.51}$$

$$\tilde{F}_p(3) = \Gamma(\alpha_{123},\, \alpha_{013},\, \alpha_{023})F_n(3), \tag{4.52}$$

$$\tilde{F}_p(4) = \Gamma(\alpha_{124},\, \alpha_{014},\, \alpha_{024})F_n(4), \tag{4.53}$$

and

$$\tilde{F}_p(5) = \Gamma(\alpha_{125},\, \alpha_{015},\, \alpha_{025})F_n(5). \tag{4.54}$$

In general, for any $\alpha_{lmn} = -N$, the relations among the 18 terminating series would be:

$$\tilde{F}_p(i) = \tilde{F}_p(j) = \tilde{F}_p(k) \tag{4.55}$$

$$= (-1)^N \tilde{F}_n(l) = (-1)^N \tilde{F}_n(m) = (-1)^N \tilde{F}_n(n). \tag{4.56}$$

Of the three generators g_1, g_2, g_3 for G_T, in the text, for the generator g_1, the 5×5 matrix representing the Weber–Erdelyi transformation (4.18) was chosen. The 72 elements of the 5×5 representation for G_T can also be generated if g_1 is any one of the matrices representing the transformation (4.22)–(4.27) or (4.35). However, if for g_1, the 5×5 unit matrix representing (4.33) is chosen, then it would result in a 4-element subgroup of G_T. Similarly, choosing (4.28) for g_1 results in an 8-element subgroup of G_T; choosing (4.19), (4.20), (4.31), or (4.32) for g_1 results in 12-element subgroups of G_T; and choosing (4.21), (4.30), (4.33), or (4.34) results in 36 element subgroups of the group G_T.

When $c = \alpha_{345}$, $-N$ determines the termination of the ${}_3F_2$ series from the definition (4.8) for F_p. It follows that (m, n) can take only the three values (3,4), (3,5), or (4,5). Since any one of the numerator variables of $F_p(l)$—viz α_{imn}, α_{jmn}, α_{kmn}—can be α_{345}, the indices i, j, k are restricted to 5, 4, or 3, which in turn implies that l can be only 0, 1, or 2.

Therefore, (m, n) being any two of 3, 4, 5 (${}^3C_2$) and l being any one of 0, 1, 2 (${}^3C_1$), it is obvious that α_{345} can occur as a numerator parameter in only the (${}^3C_1 \times {}^3C_2 =$) 9 series.

When r_i is replaced by $-r_i$, instead of the $F_p(l)$ series, the $F_n(l)$ series arise. From the definition (4.9) for the $F_n(l)$ series, (j, k), (i, k), or (i, j) can take the values (3,4), (3,5), or (4,5); so that l can be 5, 4, or 3 (${}^3C_1$) and (m, n) can be only (0,1), (0,2), or (1,2). Once again there are only nine F_n series. This explains why in the relations (4.18)–(4.22) amongst the 18 terminating ${}_3F_2$ series,

$$F_p(0),\ F_p(1),\ F_p(2)\quad \text{and}\quad F_n(3),\ F_n(4),\ F_n(5)$$

alone occur.

Along with g_1 defined in (4.15), let g_2 be the matrix

$$g_2 = \begin{bmatrix} 0 & 1 & 0 & 0 & 0 \\ 1 & 0 & 0 & 0 & 0 \\ 0 & 0 & 1 & 0 & 0 \\ 0 & 0 & 0 & 1 & 0 \\ 0 & 0 & 0 & 0 & 1 \end{bmatrix}, \tag{4.57}$$

which interchanges a and b when it operates on $\vec{x}$ and denotes g_3 the matrix:

$$g_3 = \begin{bmatrix} 1 & 0 & 0 & 0 & 0 \\ 0 & 1 & 0 & 0 & 0 \\ 0 & 0 & 1 & 0 & 0 \\ 0 & 0 & 0 & 0 & 1 \\ 0 & 0 & 0 & 1 & 0 \end{bmatrix}, \tag{4.58}$$

which interchanges d and e when it operates on $\vec{x}$. By forming all possible products of all possible powers of g_1, g_2, and g_3, a group of 72 transformation matrices can be generated, which provides a 5×5 representation for the terminating series, with (4.14) as the basis. Thus, g_1, g_2, and g_3 are the generators of a group G_T for the transformations of a terminating ${}_3F_2$ series, with $g_i^2 = \mathbf{1}$, for $i = 1,2,3$.

A similarity transformation, $u^{-1} g_i\, u$, with a

$$u = \begin{bmatrix} 1 & 0 & 1 & 0 & 0 \\ 0 & 1 & 1 & 0 & 0 \\ 0 & 0 & 3 & 0 & 0 \\ 0 & 0 & 2 & 1 & 0 \\ 0 & 0 & 2 & 0 & 1 \end{bmatrix} \quad \text{and} \quad u^{-1} = \frac{1}{3}\begin{bmatrix} 3 & 0 & -1 & 0 & 0 \\ 0 & 3 & -1 & 0 & 0 \\ 0 & 0 & 1 & 0 & 0 \\ 0 & 0 & -2 & 3 & 0 \\ 0 & 0 & -2 & 0 & 3 \end{bmatrix} \tag{4.59}$$

block diagonalizes the generators, and hence all the $g \in G_T$, thereby reducing the generators for the 5×5 representation into the generators for a one-dimensional identity irrep (due to $-N$ being kept fixed in (4.11)), and the generators for a four-dimensional faithful irrep given by:

$$\begin{bmatrix} 1 & 0 & 0 & 0 \\ 0 & -1 & 0 & 1 \\ 1 & 0 & -1 & 0 \\ 0 & 0 & 0 & 1 \end{bmatrix}, \begin{bmatrix} 0 & 1 & 0 & 0 \\ 1 & 0 & 0 & 0 \\ 0 & 0 & 1 & 0 \\ 0 & 0 & 0 & 1 \end{bmatrix} \text{ and } \begin{bmatrix} 1 & 0 & 0 & 0 \\ 0 & 1 & 0 & 0 \\ 0 & 0 & 0 & 1 \\ 0 & 0 & 1 & 0 \end{bmatrix}. \tag{4.60}$$

In terms of Whipple's parameters and the definitions of the F_p and F_n series given by (4.8) and (4.9), respectively, the transformation (4.11) can be written as:

$$F_p(0; 45) = (-1)^N \frac{\Gamma(\alpha_{015}, \alpha_{025})}{\Gamma(\alpha_{123}, \alpha_{124})} F_n(5; 02), \tag{4.61}$$

where $\alpha_{345} = -N$. (See appendix and equation (4.3.3.6) in Slater 1966.) In the Whipple parameter basis, where

$$\vec{x}' = (r_0, r_1, r_2, r_3, r_4, r_5) \tag{4.62}$$

is represented as a column vector, the transformation (4.61) is equivalent to the 6×6 transformation matrix:

$$g_1' = \begin{bmatrix} 0 & 0 & 0 & 0 & 0 & -1 \\ 0 & 0 & 0 & -1 & 0 & 0 \\ 0 & 0 & 0 & 0 & -1 & 0 \\ 0 & -1 & 0 & 0 & 0 & 0 \\ 0 & 0 & -1 & 0 & 0 & 0 \\ -1 & 0 & 0 & 0 & 0 & 0 \end{bmatrix}. \tag{4.63}$$

The permutation of the two numerator parameters a and b in the ${}_3F_2$, in terms of Whipple parameters is equivalent to an interchange of r_1 and r_2, which is induced by the matrix:

$$g_2' = \begin{bmatrix} 1 & 0 & 0 & 0 & 0 & 0 \\ 0 & 0 & 1 & 0 & 0 & 0 \\ 0 & 1 & 0 & 0 & 0 & 0 \\ 0 & 0 & 0 & 1 & 0 & 0 \\ 0 & 0 & 0 & 0 & 1 & 0 \\ 0 & 0 & 0 & 0 & 0 & 1 \end{bmatrix}, \tag{4.64}$$

operating on the basis vector $\vec{x}'$. Similarly, the permutation of the two denominator parameters d and e in the ${}_3F_2$, is equivalent to the interchange of r_4 and r_5, induced by:

$$g_3' = \begin{bmatrix} 1 & 0 & 0 & 0 & 0 & 0 \\ 0 & 1 & 0 & 0 & 0 & 0 \\ 0 & 0 & 1 & 0 & 0 & 0 \\ 0 & 0 & 0 & 1 & 0 & 0 \\ 0 & 0 & 0 & 0 & 0 & 1 \\ 0 & 0 & 0 & 0 & 1 & 0 \end{bmatrix}. \tag{4.65}$$

These three 6×6 matrices generate a six-dimensional reducible representation for G_T.

This six-dimensional representation, in the Whipple parameter basis, $\vec{x}'$, can be reduced by the similarity transformation, $u'^{-1}g_i'u'$, with:

$$u' = \begin{bmatrix} 1 & 1 & 0 & 0 & 0 & 1 \\ 1 & -1 & 1 & 0 & 0 & 1 \\ 1 & 0 & -1 & 0 & 0 & 1 \\ 1 & 0 & 0 & 1 & 0 & -1 \\ 1 & 0 & 0 & -1 & 1 & -1 \\ 1 & 0 & 0 & 0 & -1 & -1 \end{bmatrix}, \tag{4.66}$$

and

$$u'^{-1} = \frac{1}{6}\begin{bmatrix} 1 & 1 & 1 & 1 & 1 & 1 \\ 4 & -2 & -2 & 0 & 0 & 0 \\ 2 & 2 & -4 & 0 & 0 & 0 \\ 0 & 0 & 0 & 4 & -2 & -2 \\ 0 & 0 & 0 & 2 & 2 & -4 \\ 1 & 1 & 1 & -1 & -1 & -1 \end{bmatrix}, \tag{4.67}$$

which block diagonalizes the generators g_1', g_2' and g_3', and hence all of $g' \in G_T$. It results in two one-dimensional irreps, one of which is the identity irrep, and a four-dimensional faithful irrep with generators:

$$\begin{bmatrix} 0 & 0 & 0 & 1 \\ 0 & 0 & -1 & 1 \\ 1 & -1 & 0 & 0 \\ 1 & 0 & 0 & 0 \end{bmatrix}, \begin{bmatrix} 1 & 0 & 0 & 0 \\ 1 & -1 & 0 & 0 \\ 0 & 0 & 1 & 0 \\ 0 & 0 & 0 & 1 \end{bmatrix} \text{ and } \begin{bmatrix} 1 & 0 & 0 & 0 \\ 0 & 1 & 0 & 0 \\ 0 & 0 & 1 & 0 \\ 0 & 0 & 1 & -1 \end{bmatrix}. \tag{4.68}$$

From (4.63)–(4.65) it follows that G_T is a subgroup of the permutation group S_6. Indeed, the generators g_i' of G_T can be represented by 6×6 permutation matrices (including an overall minus-sign for g_1'). If we use the cycle notation for an element of S_6 represented by a 6×6 permutation matrix, we see from (4.63)–(4.65) that

$$\begin{aligned} g_1' &= -(05)(13)(24), \\ g_2' &= (12), \\ g_3' &= (45), \end{aligned} \tag{4.69}$$

where a minus sign for g_1' is included in order to remember that, in the Whipple parameter representation, this generator is actually a permutation matrix multiplied by -1. In the following section it will be very useful to represent elements of G_T by means of the above cycle notation, especially for distinguishing between conjugacy classes with the same order.

4.4 Structure of G_T and its irreps

Two elements, h and h', of a group G are said to be conjugate if there exists a $g \in G$ such that $h'=ghg^{-1}$. This defines an equivalence relation on G, the equivalence classes being called the conjugacy classes. Analysis of G_T, reveals that there are nine conjugacy classes $K_1, \ldots, K_9$. A conjugacy class is represented by one of its elements. In the following table is the list of all the conjugacy classes K_i, a representative element (given in terms of the generators, and as a permutation matrix in cycle notation), the order k_i of K_i (i.e. the number of elements of K_i), and the order of the elements of K_i (i.e. the smallest integer s such that $g^s = \mathbf{1}$, for $g \in K_i$).

Class	Order k_i of K_i	Order of $g \in K_i$	Representative of $g \in K_i$	
K_1	1	0	$g_1^2 = \mathbf{1}$	$\mathbf{1}$
K_2	4	3	$g_1g_2g_1g_3$	(345)
K_3	4	3	$(g_2g_1g_3)^2$	(012)(345)
K_4	6	2	g_1	–(05)(13)(24)
K_5	6	2	g_2	(12)
K_6	9	2	g_2g_3	(12)(45)
K_7	12	6	$g_2g_1g_3$	–(051324)
K_8	12	6	$g_1g_3g_2g_1g_2$	(021)(34)
K_9	18	4	g_1g_2	–(05)(1423)

Following the general theory of group representations (Vergados 2017 or Wybourne 1964), the table of characters for the irreps of G_T has been obtained. As there are nine conjugacy classes, there are nine inequivalent irreps, which are denoted by $D^{(1)}, \ldots, D^{(9)}$. Four irreps are of dimension 1, one is of dimension 2, and four are of dimension 4. It is only the four-dimensional irreps which are faithful. The following table lists the characters:

	K_1	K_2	K_3	K_4	K_5	K_6	K_7	K_8	K_9
$D^{(1)}$	1	1	1	1	1	1	1	1	1
$D^{(2)}$	1	1	1	−1	1	1	−1	1	−1
$D^{(3)}$	1	1	1	1	−1	1	1	−1	−1
$D^{(4)}$	1	1	1	−1	−1	1	−1	−1	1
$D^{(5)}$	2	2	2	0	0	−2	0	0	0
$D^{(6)}$	4	1	−2	0	2	0	0	−1	0
$D^{(7)}$	4	1	−2	0	−2	0	0	1	0
$D^{(8)}$	4	−2	1	2	0	0	−1	0	0
$D^{(9)}$	4	−2	1	−2	0	0	1	0	0

by looking at the traces of g_1 and g_2, and comparing them with the columns K_4 and K_5 (of which g_1 and g_2 are representatives) in the character table, it is possible to conclude that the representation generated by g_i ($i = 1,2,3$) is equivalent to

$$D^{(1)} \oplus D^{(6)}, \tag{4.70}$$

and that the Whipple parameter representation generated by g_i', $i = 1, 2, 3$ is equivalent to

$$D^{(1)} \oplus D^{(2)} \oplus D^{(6)}. \tag{4.71}$$

As a consequence, the irreducible representation matrices (4.15), (4.57), and (4.58) for the generators of G_T are equivalent and both can be labeled with $D^{(6)}$.

The next property to analyse is the simplicity of G_T. All the invariant subgroups H of G_T have been found. Among these there are proper abelian invariant subgroups, hence G_T is neither simple nor semi-simple. Recall that a subgroup H is an invariant subgroup (self-conjugate subgroup, normal divisor) if $G_T H G_T^{-1} = H$. To find invariant subgroups, one can form unions of conjugacy classes and check if they close under the group multiplication law. The following inclusion table gives a complete list of the invariant subgroups of G_T (the subscript denoting the order of H):

$$H_9 \subset H_{18} \begin{cases} \subset H_{36} \subset G_T, \\ \subset H'_{36} \subset G_T, \\ \subset H''_{36} \subset G_T, \end{cases} \tag{4.72}$$

where

$$\begin{aligned} H_9 &= K_1 \bigcup K_2 \bigcup K_3, \\ H_{18} &= H_9 \bigcup K_6, \\ H_{36} &= H_{18} \bigcup K_9, \\ H'_{36} &= H_{18} \bigcup K_4 \bigcup K_7, \\ H''_{36} &= H_{18} \bigcup K_5 \bigcup K_8. \end{aligned} \tag{4.73}$$

It should be noted that, in terms of the three generators g_i (or g_i') introduced previously, one can write

$$K_6 = g_2 g_3 H_9, \quad K_9 = g_1 g_2 H_{18}, \quad K_4 \bigcup K_7 = g_1 H_{18}, \quad K_5 \bigcup K_8 = g_2 H_{18}, \tag{4.74}$$

such that the invariant subgroups can be characterized as follows in terms of H_9 and the three generators:

$$\begin{aligned} H_{18} &= H_9 \bigcup g_2 g_3 H_9, \\ H_{36} &= H_9 \bigcup g_2 g_3 H_9 \bigcup g_1 g_2 H_9 \bigcup g_1 g_3 H_9, \\ H'_{36} &= H_9 \bigcup g_2 g_3 H_9 \bigcup g_1 H_9 \bigcup g_1 g_2 g_3 H_9, \\ H''_{36} &= H_9 \bigcup g_2 g_3 H_9 \bigcup g_2 H_9 \bigcup g_3 H_9. \end{aligned} \tag{4.75}$$

The smallest invariant subgroup, H_9, is easy to characterize. In fact,

$$H_9 = C_3 \times C_3, \tag{4.76}$$

the direct product of two cyclic groups on three elements. In terms of the Whipple parametrization, the generators of the two C_3's are (012) and (345). It is now obvious that H_9 is an abelian invariant subgroup of G_T.

It should be noticed that all the invariant subgroups of G_T can be found using the character table, and that those elements h of G_T with $\phi(h) = \phi(\mathbf{1})$—where ϕ is a (not necessarily simple) character of G_T—form an invariant subgroup (Ledermann 1977, theorem 2.7)

Conversely, having the list of all invariant subgroups of G_T, one can reconstruct the character table. Indeed, the first character $\chi^{(1)}$ is trivial. Next, if N is one of H_{36}, H'_{36}, or H''_{36}, G/N is the two-element group C_2, with a non-trivial simple character $(1, -1)$. Using the 'lifting process' (Ledermann 1977, theorem 2.6), one obtains the simple characters $\chi^{(2)}$, $\chi^{(3)}$, and $\chi^{(4)}$ from H''_{36}, H'_{36}, and H_{36}, respectively. This completes the list of simple characters with $\chi_1^{(i)} = 1$. In order to find the remaining simple characters, the theory of induced characters can be used. If H is a subgroup of G for which a character ${}^H\phi$ is known, then

$$ {}^G\phi_i = \frac{m}{k_i} \sum_w {}^H\phi(w), \qquad w \in K_i \cap H \tag{4.77} $$

is a character (simple or compound) of G. Herein, m is the index of H and k_i is the order of K_i. As the simple characters of an abelian group are well known, H is chosen to be $H_9 = C_3 \times C_3$, thus $m = 72/9 = 8$. Using the trivial character of H, ${}^H\phi^{(1)} = (1,1,1,1,1,1,1,1,1)$, one finds ${}^G\phi^{(1)} = (8,8,8,0,0,0,0,0,0)$. By means of the inner product for the characters of G_T,

$$ \langle \phi | \psi \rangle = \frac{1}{72} \sum_{i=1}^{9} k_i \phi_i \psi_i, \tag{4.78} $$

it is found that

$$ \langle {}^G\phi^{(1)} | \chi^{(1)} \rangle = \langle {}^G\phi^{(1)} | \chi^{(2)} \rangle = \langle {}^G\phi^{(1)} | \chi^{(3)} \rangle = \langle {}^G\phi^{(1)} | \chi^{(4)} \rangle = 1. \tag{4.79} $$

Thus, subtracting $\chi^{(1)}, \ldots, \chi^{(4)}$ from ${}^G\phi^{(1)}$, one obtains

$$ {}^G\phi' = (4, 4, 4, 0, 0, -4, 0, 0, 0). \tag{4.80} $$

Since all one-dimensional irreps have been found and

$$ \langle {}^G\phi' | {}^G\phi' \rangle = 4, \tag{4.81} $$

it follows that ${}^G\phi'$ is twice a simple character, i.e. ${}^G\phi' = 2\chi^{(5)}$. The next simple character, $\chi^{(6)}$, is immediately deduced from our defining representation (4.14), (4.15), (4.57) and (4.58). Using a non-trivial character of H,

$$ {}^H\phi^{(2)} = (1, 1, 1, \omega, \omega, \omega, \omega^2, \omega^2, \omega^2), \tag{4.82} $$

where

$$ \omega^2 + \omega + 1 = 0, \tag{4.83} $$

the inducing process leads to

$$ {}^G\phi^{(2)} = (8, 2, -4, 0, 0, 0, 0, 0, 0). \tag{4.84} $$

One can verify that the inner product of ${}^{G}\phi^{(2)}$ with $\chi^{(1)}, \chi^{(2)}, \chi^{(3)}, \chi^{(4)}$ and $\chi^{(5)}$ is zero, and that

$$\langle {}^{G}\phi^{(2)} | \chi^{(6)} \rangle = 1. \tag{4.85}$$

Subtracting $\chi^{(6)}$ from ${}^{G}\phi^{(2)}$, one obtains

$${}^{G}\phi'' = (4, 1, -2, 0, -2, 0, 0, 1, 0). \tag{4.86}$$

Since

$$\langle {}^{G}\phi'' | {}^{G}\phi'' \rangle = 1, \tag{4.87}$$

it is a simple character, that is

$${}^{G}\phi'' = \chi^{(7)}. \tag{4.88}$$

Two more simple characters $\chi^{(8)}$ and $\chi^{(9)}$ need to be found. Using the orthogonality property satisfied by the columns of the character table of G_T, namely

$$\sum_{l=1}^{9} \chi_i^{(l)} \chi_j^{(l)} = \frac{72}{k_i} \delta_{ij}, \tag{4.89}$$

it is a straightforward exercise to complete the character table.

Although in the preceding sections G_T was generated by three generators, namely the Weber–Erdelyi transformation g_1 and the two interchange transformations $a \leftrightarrow b$ (g_2) and $d \leftrightarrow e$ (g_3), it should be noted that G_T can actually be generated by only two elements. For instance, using the cycle structure notation for the elements of G_T, the 72-element group G_T is generated by (4.36) and $-(0524)(31)$, i.e. by g_2 and $(g_1 g_3)$. In fact there are many other examples of pairs of generators for G_T.

4.5 Scaling the Weber–Erdelyi transformation

Using the notation of section 4.1, the Weber–Erdelyi transformation (4.11) can be written in the following form:

$${}_3F_2(\vec{x}) = \frac{\Gamma(d, d + N - a)}{\Gamma(d + N, d - a)} {}_3F_2(g_1\vec{x}), \tag{4.90}$$

whereas the interchange transformations are:

$${}_3F_2(\vec{x}) = {}_3F_2(g_2\vec{x}), \qquad {}_3F_2(\vec{x}) = {}_3F_2(g_3\vec{x}). \tag{4.91}$$

In general, this analysis implies that

$${}_3F_2(\vec{x}) = (\text{factor})\, {}_3F_2(g\vec{x}), \qquad \forall\, g \in G_T, \tag{4.92}$$

where this factor is in terms of Γ-functions, as in (4.11) or (4.16). It would be interesting if this factor could actually be determined in terms of the group element g.

This can indeed be done. The most elegant way to obtain this is to perform a scaling on the ${}_3F_2(\vec{x})$:

$$ {}_3\tilde{F}_2(\vec{x}) = \frac{\Gamma(d+N,\, e+N)}{\Gamma(d,\, e)}\, {}_3F_2(\vec{x}). \tag{4.93} $$

Then the three generating transformations become:

$$ \begin{aligned} {}_3\tilde{F}_2(\vec{x}) &= (-1)^N\, {}_3\tilde{F}_2(g_1\vec{x}), \\ {}_3\tilde{F}_2(\vec{x}) &= {}_3\tilde{F}_2(g_2\vec{x}) = {}_3\tilde{F}_2(g_3\vec{x}). \end{aligned} \tag{4.94} $$

As G_T is generated by g_1, g_2, and g_3, the following result holds: the scaled terminating ${}_3\tilde{F}_2$ with unit argument satisfies

$$ \begin{aligned} {}_3\tilde{F}_2(\vec{x}) &= {}_3\tilde{F}_2(g\vec{x}), \qquad \forall\, g \in G_T, \qquad \text{(for } N \text{ even)}, \\ {}_3\tilde{F}_2(\vec{x}) &= \chi^{(2)}(g)\, {}_3\tilde{F}_2(g\vec{x}), \qquad \forall\, g \in G_T, \qquad \text{(for } N \text{ odd)}, \end{aligned} \tag{4.95} $$

where $\chi^{(2)}(g)$ is the character of g in the irrep $D^{(2)}$ (see section 4.4). Hence the 72-element group G_T can be seen as the invariance group of the terminating ${}_3F_2$. If N is odd, then the coefficient in (32) is +1 or −1, and it is equal to −1 if one of the following equivalent conditions is satisfied:

- g_1 appears an odd number of times in the expression of g in terms of g_1, g_2 and g_3;
- g is a permutation matrix times −1 when represented in the Whipple parametrization;
- the left and right-hand sides of (4.32) correspond to a F_p and a F_n in terms of the notation of section 4.2.

In the next chapter, we show how this transformation of Weber–Erdelyi can be used to obtain from the highly symmetric Vander Waerden form of the 3-j coefficient the three other forms discovered by Wigner, Racah, and Majumdar, by different methods.

4.6 Non-terminating ${}_3F_2(a, b, c; d, e; 1)$ series

The two-term relation for the non-terminating ${}_3F_2$ series of the unit argument given by Thomae (1879) is

$$ \begin{aligned} {}_3F_2(a, b, c; d, e; 1) &\equiv {}_3F_2(a, b, c; d, e) \\ &= \frac{\Gamma(d, e, s)}{\Gamma(a, s+b, s+c)}{}_3 \\ &\qquad F_2(d-a, e-a, s; s+b, s+c) \end{aligned} \tag{4.96} $$

where, for simplicity of notation, the unit argument feature of the ${}_3F_2(1)$ will be tacit. As in the case of the terminating ${}_3F_2$ series, a recursive use of (4.86) was shown by

Srinivasa Rao *et al* (1998) to result in a set of ten non-terminating transformations given below:

$$ {}_3F_2(a, b, c; d, e) = \frac{\Gamma(d, e, s)}{(a, s+b, s+c)} \times {}_3F_2(d-a, e-a, s; s+b, s+c) \tag{4.97} $$

$$ = \frac{\Gamma(d, s)}{(d-a, s+a)} {}_3F_2(a, e-b, e-c; e, s+a) \tag{4.98} $$

$$ = \frac{\Gamma(e, s)}{(e-c, s+c)} {}_3F_2(d-a, d-b, c; s+c, d) \tag{4.99} $$

$$ = \frac{\Gamma(d, e, s)}{(s+a, s+b, c)} {}_3F_2(s, d-c, e-c; s+a, s+b) \tag{4.100} $$

$$ = {}_3F_2(a, b, c, ;d, e) \qquad \text{(identity)} \tag{4.101} $$

$$ = \frac{\Gamma(e, s)}{(s+a, e-a)} {}_3F_2(a, d-b, d-c; s+a, d) \tag{4.102} $$

$$ = \frac{\Gamma(d, s)}{(d-c, s+c)} {}_3F_2(e-a, e-b, c; s+a, e) \tag{4.103} $$

$$ = \frac{\Gamma(e, s)}{(e-b, s+b)} {}_3F_2(d-a, b, d-c; d, s+b) \tag{4.104} $$

$$ = \frac{\Gamma(d, e, s)}{(s+a, b, s+c)} {}_3F_2(d-b, e-b, s; s+a, s+c) \tag{4.105} $$

$$ = \frac{\Gamma(d, s)}{(d-b, s+b)} {}_3F_2(e-a, b, e-c; s+b, e). \tag{4.106} $$

Let us denote the parameters of the ${}_3F_2$ on the right-hand side and the left-hand side of the Thomae transformation (4.96) by the column vectors

$$ \vec{x} = (a, b, c, d, e) \quad \text{and} \quad \vec{x'} = (a', b', c', d', e'). \tag{4.107} $$

They are then related by the linear Thomae transformation

$$\vec{x}' = g_1\vec{x}, \tag{4.108}$$

where the transformation matrix g_1 is

$$g_1 = \begin{pmatrix} -1 & 0 & 0 & 1 & 0 \\ -1 & 0 & 0 & 0 & 1 \\ -1 & -1 & -1 & 1 & 1 \\ -1 & 0 & -1 & 1 & 1 \\ -1 & -1 & 0 & 1 & 1 \end{pmatrix}. \tag{4.109}$$

If on these ten transformations we superpose the manifest permutations of the numerator and denominator variables, then we get a set of 120 elements. The 5×5 matrix representation for this set can be obtained by taking into account, in addition to the generator g_1, the generators of the S_3 and S_2 groups:

$$g_2 = \begin{bmatrix} 0 & 1 & 0 & 0 & 0 \\ 0 & 0 & 1 & 0 & 1 \\ 1 & 0 & 0 & 0 & 0 \\ 0 & 0 & 0 & 1 & 0 \\ 0 & 0 & 0 & 0 & 1 \end{bmatrix}, \quad g_3 = \begin{bmatrix} 0 & 1 & 0 & 0 & 0 \\ 1 & 0 & 0 & 0 & 0 \\ 0 & 0 & 1 & 0 & 0 \\ 0 & 0 & 0 & 1 & 0 \\ 0 & 0 & 0 & 0 & 1 \end{bmatrix}$$

and

$$g_4 = \begin{bmatrix} 1 & 0 & 0 & 0 & 0 \\ 0 & 1 & 0 & 0 & 0 \\ 0 & 0 & 1 & 0 & 0 \\ 0 & 0 & 0 & 0 & 1 \\ 0 & 0 & 0 & 1 & 0 \end{bmatrix}. \tag{4.110}$$

A scaling on the ${}_3F_2(\vec{x})$ through the transformation

$${}_3\tilde{F}_2(\vec{x}) = \frac{1}{\Gamma(d, e, s)}\,{}_3F_2(\vec{x}) \tag{4.111}$$

enables us to write the Thomae transformation (4.106) as

$${}_3\tilde{F}_2(\vec{x}) = {}_3\tilde{F}_2(g_1\vec{x}). \tag{4.112}$$

In general, this implies that the scaled non-terminating ${}_3\tilde{F}_2(1)$ satisfies

$${}_3\tilde{F}_2(\vec{x}) = {}_3\tilde{F}_2(g\,\vec{x}), \quad \forall \quad g\ \varepsilon\ G_T, \tag{4.113}$$

where G_T is the 120 element invariance group of the non-terminating ${}_3F_2$ transformations. That this invariance group is the symmetric group S_5 has been proved by Beyer *et al* (1987).

4.7 Concluding remarks

It is well known that one of the van der Waerden forms for the 3-j coefficient can be written as follows:

$$\begin{pmatrix} j_1 & j_2 & j_3 \\ m_1 & m_2 & m_3 \end{pmatrix} = \delta(m_1 + m_2 + m_3, 0)(-1)^{j_1 - j_2 - m_3}$$
$$\times [(-j_1 + j_2 + j_3)!(j_1 - j_2 + j_3)!(j_2 - m_2)!(j_3 - m_3)!(j_1 + m_1)!(j_3 + m_3)!]^{1/2}$$
$$\times [(j_1 + j_2 - j_3)!(j_1 + j_2 + j_3 + 1)!(j_1 - m_1)!(j_2 + m_2)!]^{-1/2} \tag{4.114}$$
$$\times [(j_3 - j_1 - m_2)!(j_3 - j_2 + m_1)!]^{-1}$$
$$\times\, {}_3F_2\left(\begin{matrix} -j_1 + m_1, & -j_2 - m_2, & -j_1 - j_2 + j_3 \\ 1 + j_3 - j_1 - m_2, & & 1 + j_3 - j_2 + m_1 \end{matrix} ; 1 \right).$$

The three generating elements g_1, g_2, and g_3 of G_T lead, respectively, to the following symmetries of the 3-j symbol (apart from a phase factor):

$$\begin{pmatrix} j_1 & -j_3 - 1 & -j_2 - 1 \\ m_1 & m_3 & m_2 \end{pmatrix}, \tag{4.115}$$

$$\begin{pmatrix} (j_1 + j_2 - m_3)/2 & (j_1 + j_2 + m_3)/2 & j_3 \\ (j_1 - j_2 + m_1 - m_2)/2 & (j_1 - j_2 - m_1 + m_2)/2 & -j_1 + j_2 \end{pmatrix}, \tag{4.116}$$

$$\begin{pmatrix} (j_1 + j_2 + m_3)/2 & (j_1 + j_2 - m_3)/2 & j_3 \\ (-j_1 + j_2 + m_1 - m_2)/2 & (-j_1 + j_2 - m_1 + m_2)/2 & j_1 - j_2 \end{pmatrix}. \tag{4.117}$$

The second and third of these are well known Regge symmetries (1958) of the 3-j symbol, while the first has unphysical arguments (the j-values being negative; the triangular condition is violated). The classical symmetry group of the 3-j coefficient contains 72 symmetries, of which (4.116) and (4.117) are two elements. Following Louck *et al* (1986), who extended the classical Regge group of 144 symmetries of the 6-j symbol by the ${}_4F_3$ invariance group S_5 in order to obtain a new symmetry group of the order 23040, one can perform the same process here and extend the 72 classical symmetries of the 3-j symbol to the symmetries induced by the 72-element group G_T. Since (4.116) and (4.117) are Regge symmetries, already contained in the 72 symmetries; this amounts to the enlarging of these symmetries by the element (4.115) and to investigating which group G it generates. In particular, (4.115) contains unphysical transformations of the type $j \to -j - 1$ (preserving the angular momentum eigenvalue $j(j + 1)$), known as Yutsis mirror symmetries (Yutsis *et al* 1960). Let us denote $j_1 \to -j_1 - 1$ by r'. It can be shown by recursively using r' and the column permutations of the 3-j coefficient that (4.115) can be transformed into

$$\begin{pmatrix} -j_1 - 1 & j_2 & j_3 \\ m_1 & m_2 & m_3 \end{pmatrix}. \tag{4.118}$$

The group G can be generated by the classical symmetries together with r'. This new group G is of order 1440; it can be interpreted as the extended symmetry group of the 3-j coefficient by extending the domain of this coefficient. This extended domain contains unphysical arguments. It should be noticed that this extended symmetry group of the order 1440 has been encountered by D'Adda *et al* (1971) in treating $SU(2)$ and $SU(1,1)$ 3-j coefficients, and by Huszár (1972). There are two further observations to make. The first is that the 'trivial' ${}_3F_2$ symmetry permuting two of the three numerator parameters corresponds to a non-trivial Regge (1958) symmetry for the 3-j symbol (in fact, this observation is not new: see Biedenharn and Louck 1984, p 433). The second, new, observation is that a 'trivial' 3-j symmetry (namely $j_1 \to -j_1 - 1$) corresponds to a non-trivial transformation for the terminating ${}_3F_2(1)$ series, namely to (4.11).

It is considered relevant to point out the work of Beyer *et al* (1987) in the present context. For this purpose, in the Whipple notation (section 4.2) let ℓ, m, n be 0, 4, 5, respectively. Then the numerator and denominator variables, which occur in $F_p(0; 45)$, given by (4.8), after elimination of r_0 using (4.6), are related to the five independent Whipple parameters:

$$\vec{r} = (r_1, r_2, r_3, r_4, r_5) \tag{4.119}$$

through the transformation:

$$\vec{\alpha} = A\vec{r}, \tag{4.120}$$

where

$$\vec{\alpha} = \left(\alpha_{145} - \frac{1}{2}, \alpha_{245} - \frac{1}{2}, \alpha_{345} - \frac{1}{2}, \beta_{40} - 1, \beta_{50} - 1\right), \tag{4.121}$$

and

$$A = \begin{bmatrix} 1 & 0 & 0 & 1 & 1 \\ 0 & 1 & 0 & 1 & 1 \\ 0 & 0 & 1 & 1 & 1 \\ 1 & 1 & 1 & 2 & 1 \\ 1 & 1 & 1 & 1 & 2 \end{bmatrix}. \tag{4.122}$$

This 5×5 matrix A plays a crucial role in the study of the group structure of the two-term identities by Beyer *et al* (1987). They analyse the group structure of the non-terminating series and establish that the symmetric group S_5 is an invariance group of the two-term relation for the ${}_3F_2$ series (Thomae (1879)) and the invariance of that series to separate permutations of the numerator and denominator parameters of the ${}_3F_2$.

We generated a 72-element group G_T for the terminating ${}_3F_2(1)$ series, presented the conjugacy classes, irreps, their characters, the invariant subgroups of G_T, and discussed the role of these terminating series for the ${}_3F_2(1)$ forms of the 3-*j* coefficient. The group G_T, which is of interest to us, has been reached by the simple recursive use of a given ${}_3F_3(1)$ transformation, and the results presented for the terminating ${}_3F_2(1)$ series supplement the work of Beyer *et al* (1985). The structure of the invariance group G_T for the terminating ${}_3F_2(1)$ series has turned out to be more intricate than that of the symmetric group S_5 shown to be the invariance group for the non-terminating ${}_3F_2(1)$ series investigated by Beyer *et al* (1985). Our study contributes to a complete understanding of an interesting aspect overlooked in the work of Beyer *et al* (1985).

References

Bailey W N 1935 Generalized Hypergeometric Series *Cambridge Tracts in Mathematics and Mathematical Physics* 32 (Cambridge: Cambridge University Press)

Beyer W A, Louck J D and Stein P R 1985 Preprint, LA-UR-85-3561

Beyer W A, Louck J D and Stein P R 1987 *J. Math. Phys.* **28** 497

Biedenharn L C and Louck J D 1984 The Racah Wigner Algebra in Quantum Theory *Encyclopedia of Mathematics and its Applications* **vol 9**

D'Adda A, D'Auria R and Ponzano G 1971 On generalized Wigner and Racah Coefficients *Internal report* September 1971.

D'Adda A, D'Auria R and Ponzano G Also, see 1972 *Proc. of the Int. School of Physics 'Enrico Fermi'* vol 54

Huszár M 1972 *Acta Phys. Acad. Sci.* **32** 185

Ledermann W 1977 *Introduction to Group Characters* 2nd edition (Cambridge: Cambridge University Press)

Louck J D, Beyer W A, Biedernharn L C and Stein P R 1986 Symmetries of some hypergeometric series: implications for 3-j and 6-j symbols *Los Alamos National Laboratory report* LA-UR-86-4203

Regge T 1958 *Nuovo Cimento* **10** 544

Slater L J 1966 *Generalized Hypergeometric Functions* (Cambridge: Cambridge University Press)

Srinivasa Rao K, Van der Jeugt J, Raynal J, Jagannathan R and Rajeswari V 1992 *J. Phys. A: Math. Gen* **25** 861–76

Srinivasa Rao K and Rajeswari V 1993 *Selected Topics in Quantum Theory of Angular Momentum* (New York: Springer)

Srinivasa Rao K, Doebner H-D and Natterman P 1998 *Generalized Hypergeometric Series and the Symmetries of 3-j and 6-j Coefficients Number Theoretic Methods: Developments in Mathematics* ed S Kanemitsu and C Jia vol 8 (New York: Springer)

Smorodinskii Y A and Shelepin Leonid A 1972 *Sov. Phys. Uspekhi* **15** 1–24

Thomae J 1879 *J. für Math* **87** 26–73

Vergados J D 2017 *Group and Representation Theory* (Singapore: World Scientific)

Weber M and Erfely A 1952 *Am. Math. Monthly* **59** 163

Whipple F J W 1925 *Proc. London Math. Soc.* **23** 104–14

Wybourne B G 1964 *Classical Groups for Physicists* (New York: Wiley)

Yutsis A P, Levinson I B and Vanagas V V 1960 *The Theory of Angular Momentum* (Vilnius: The Israel Program for Scientific Translations)

IOP Publishing

Generalized Hypergeometric Functions
Transformations and group theoretical aspects
K Srinivasa Rao and Vasudevan Lakshminarayanan

Chapter 5

Angular momentum and the rotation group

5.1 Introduction: historical

In this chapter, we introduce the classical definitions of linear and angular momentum. We show how the introduction of the quantum mechanical replacements of *vecp* by $-i\hbar\nabla$, and the introduction of the basic commutation relation for the canonically conjugate variables x and p_x, leads to the non-vanishing of the commutation relations for the components of orbital angular momentum. The foundation for the quantum theory of angular momentum—an integral part of quantum mechanics—was laid out in the 1920s, and profound theoretical developments followed.

In physics, group theory had its use only after the advent of quantum mechanics—the foundation for which was laid by Erwin Rudolf Joseph Alexander Schrödinger[1], Paul Adrian Maurice Dirac, and Werner Heisenberg.

Group theory was applied to calculations of atomic structures and spectra, first by Elliott ((1958) see also Arima 1999), in what is known as Elliott's $SU(3)$, and most notably by Eugene Paul Wigner[2] and Hans Albrecht Bethe[3]. The prediction of the Ω^- particle from an application of group theory to elementary particle physics by Yuval Neéman[4] and Murray Gell-Mann[5] established group theory as an essential, indispensable tool in physics and physical chemistry.

One of the greatest achievements of Neéman in physics was his 1961 discovery of the classification of hadrons through the $SU(3)$ flavour symmetry, now named the eightfold way, which was also proposed independently by Murray Gell-Mann (Gellman and Neeman, 1964). This $SU(3)$ symmetry laid the foundation for the quark model, proposed by Gell-Mann and George Zweig in 1964 (independently of

[1] Erwin Schrodinger (1887–1961), an Austrian physicist, shared a Nobel prize with Dirac, see Moore (2017).
[2] Eugene Paul Wigner (1902–95), Nobelist, American physicist. See Stanton and Wigner, (1995)
[3] Hans Albrecht Bethe (1906–2005), Nobelist, American physicist. See Brown and Lee (2006).
[4] Yuval Neéman (1925–2006), Israeili physicist and politician. See Joffe (2006).
[5] S Murray Gell-Mann (1929), Nobelist, American physicist. See Johnson (1999).

doi:10.1088/978-0-7503-1496-1ch5

each other). Zweig proposed the existence of quarks at CERN, Geneva, independently of Murray Gell-Mann, right after defending his PhD dissertation. Zweig dubbed them 'aces', after the four playing cards, because he speculated there were four of them (on the basis of the four extant leptons known at the time, see Zweig[6] 1980). The introduction of quarks provided a cornerstone for particle physics. A popular account of the history of particle physics is given in Crease and Mann (1996).

The collection of reprints and original papers in the anthology, *Quantum Theory of Angular Momentum* (Biedenharn and Van Dam 1965, see also Biedenharn *et al* 1952), traces the development of this theory from its origins in atomic and nuclear spectroscopy. There are a number of standard textbooks that provide the essential framework, starting from the classical concept of orbital angular momentum, $\vec{L} = \vec{r} \times \vec{p}$, Heisenberg's uncertainty principle, $\Delta x\ \Delta p_x\ \ \hbar$, and the resultant commutation relations for angular momentum—through the coupling and recoupling of two and three angular momenta, tensor operators, and the maze of formulae essential for applications to atomic, molecular, and nuclear structure and spectra. Representatives of this genre, primarily concerned with the quantum theory of angular momentum, are the classic works of Born and Jordan (1925), Rose (1955, 1957), Edmonds (1957), Fano and Racah (1959), Yutsis *et al* (1960), Brink and Satchler (1962), Varshalovich *et al* (1975), and Zare (1988), amongst others.

The major fields of application of the powerful techniques of the quantum theory of angular momentum are in several branches of physics. Condon and Shortley (1935) provided the first standard text in the class, called *Theory of Atomic Spectra*; the others are by Mayer and Jensen (1955), Feenberg (1955), Elliott and Lane (1957), Amos de-Shalit and Talmi (1953), Bohr and Mottelson (1969), and Eisenberg and Greiner (1975, 1976).

Eugene Paul Wigner (1927) is credited with the first application of group theory to analyze the significance of *rotational invariance* for atomic spectroscopy. Wigner defined and discussed the importance of rotational matrices. The Wigner 3-j coefficients and the Wigner–Eckart[7] theorem—which, in the words of Giulio Racah[8], separates the *physical aspects* of the matrix element of a tensor operator from the purely *geometric aspects*—are today part and parcel of the theory of angular momentum.

Hermann Weyl[9] was the first to publish a book on group theory and quantum mechanics, in 1928 (see Weyl 2014). It was followed by other books on group theory, of particular interest to the physicists are those of Wigner (1959), Gel'fand, Minlos and Shapiro (1958), and Hamermesh (1962), amongst others.

The celebrated four papers of Giulio Racah (1942a, 1942b, 1943, 1949) on the *Theory of Complex Spectra* are path-breakers, in which the algebraic techniques are systematically developed, and invaluable tools for the theorists are provided. Racah introduced a coefficient, since named after him, first as the recoupling scheme for

---

[6] G George Zweig (1937), American physicist.

[7] Carl Henry Eckart (1902–73), American physicist.

[8] Giulio Racah (1909–65), Italian Israeli physicist.

[9] Hermann Klaus Hugo Weyl (1885–1955), German American mathematician. See: http://www.nasonline.org/publications/biographical-memoirs/memoir-pdfs/weyl-hermann.pdf.

evaluating the scalar product of tensor operators, and later as a transformation coefficient between alternate coupling schemes for the addition of three angular momenta. He also introduced the seniority quantum number and the coefficient of fractional parentage.

Julian Schwinger[10], in 1952, developed the entire quantum theory of angular momentum from the framework of second quantized boson systems. Though this work was then unpublished, it has been reprinted as an article in Biedenharn and Van Dam (1965), which continues to inspire researchers.

With this background, Srinivasa Rao and Rajeswari (1993), in a research monograph, presented their studies of the following.

- The connection between angular momentum coefficients and generalized hypergeometric functions of the unit argument.
- The transformation theory of generalized hypergeometric functions and the different forms that exist for an angular momentum coefficient.
- Relations between angular momentum coefficients and orthogonal polynomials.
- The polynomial (or, *non-trivial*) zeros of angular momentum coefficients.
- Numerical algorithms for the computation of angular momentum coefficients based on the sets of generalized hypergeometric coefficients and the algorithms for generating the polynomial zeros.

In this chapter, in section 5.2, we present the required angular momentum algebra; in section 5.3, the two- and three-dimensional representations of the angular momentum operators; in section 5.4 the rotation group; in section 5.5, the use of the Thomae transformation to relate the different forms of the ${}_3F_2(1)$ for the angular momentum coupling coefficient; and in section 5.6, the sets of ${}_4F_3(1)$s and their relationship to each other through the property of reversal of series of the ${}_4F_3(1)$.

5.2 Angular momentum algebra

A particle of mass m and velocity $\vec{v}$, located at a point $\vec{r}$, measured from the origin of a coordinate system, has a *linear momentum*

$$\vec{p} = m\vec{v} \tag{5.1}$$

and an *angular momentum*:

$$\vec{L} = \vec{r} \times \vec{p}. \tag{5.2}$$

The classical mechanical concept becomes an operator in quantum mechanics, since the momentum is

$$\vec{p} \text{ is replaced by } -i\hbar\nabla, \quad \hbar = h/2\pi, \tag{5.3}$$

[10] Julian Seymour Schwinger (1918–1994), American physicist, Nobelist. See Mehra and Mitan (2000).

h being Planck's constant, $i = \sqrt{-1}$ and ∇ the gradient operator, which in the orthogonal Cartesian coordinate system is:

$$\nabla = \hat{i}\frac{\partial}{\partial x} + \hat{j}\frac{\partial}{\partial y} + \hat{k}\frac{\partial}{\partial z}, \tag{5.4}$$

where $\hat{i}$, $\hat{j}$, $\hat{k}$ are the unit vectors along the orthogonal Cartesian coordinate axes X, Y, Z, respectively.

Note: it is customary to use the *natural system of units*, in which

$$\hbar = c = m = 1, \text{ or the Compton wavelength, } \hbar/mc = 1, \tag{5.5}$$

of a particle of mass m is set as the unit of length throughout the calculations/computations and, in the final step, the result of the calculation/computation is multiplied by the required M, L, T (mass, length and time) dimensional units. Accordingly,

$$L_x = yp_z - zp_y = -i\left(y\frac{\partial}{\partial z} - z\frac{\partial}{\partial y}\right), \tag{5.5a}$$

$$L_y = zp_x - xp_z = -i\left(z\frac{\partial}{\partial x} - x\frac{\partial}{\partial z}\right), \tag{5.5b}$$

$$L_z = xp_y - yp_x = -i\left(x\frac{\partial}{\partial y} - y\frac{\partial}{\partial x}\right). \tag{5.5c}$$

The commutator $[A, B]$ of the two operators A, B is:

$$[A, B] = AB - BA, \tag{5.6}$$

which plays a pivotal role in quantum mechanics. The necessary and sufficient condition is that

$$[A, B] = 0. \tag{5.7}$$

Quantum mechanically, the components of the momentum operator are:

$$p_x = -i\frac{\partial}{\partial x}, \quad p_y = -i\frac{\partial}{\partial y}, \quad p_z = -i\frac{\partial}{\partial z}. \tag{5.8}$$

The component of the position operator of a particle and its corresponding momentum, satisfies the commutation relations:

$$[x, p_x] = [y, p_y] = [z, p_z] = i. \tag{5.9}$$

For,

$$
\begin{aligned}
[x, p_x]\psi(x) &= (xp_x - p_x x)\psi(x) \\
&= -i\left(x\frac{\partial}{\partial x} - \frac{\partial}{\partial x}x\right)\psi(x) \\
&= -ix\frac{\partial\psi(x)}{\partial x} + i\frac{\partial}{\partial x}(x\psi(x)) \\
&= -ix\frac{\partial\psi(x)}{\partial x} + i\psi(x) + ix\frac{\partial\psi(x)}{\partial x} \\
&= i\psi(x),
\end{aligned}
$$

and with the cancellation of the $ix\frac{\partial}{\partial x}$ term, we get the required result:

$$
[x, p_x]\psi(x) = i\psi(x) \Rightarrow [x, p_x] = i. \tag{5.10}
$$

All other commutation relations between the components of $\vec{r}$ and $\vec{p}$ are zero, i.e.

$$
[x_k, p_\ell] = i\delta_{k\ell}, \tag{5.11}
$$

where the Kronecker delta function $\delta_{k\ell}$ is defined as:

$$
\delta_{k\ell} = \begin{cases} +1, \text{ for } k = \ell \\ 0, \text{ for } k \neq \ell. \end{cases} \tag{5.12}
$$

Using this commutation relation—which is nothing but Heisenberg's uncertainty principle, (see for example, Sen 2014)—we can derive the commutation relations satisfied by the components of orbital angular momentum.

Explicitly, it follows that the position vector of a particle and its momentum satisfy the basic commutation relations:

$$
[x, p_x] = i, \quad [x, p_y] = 0 = [x, p_z], \quad \text{cyclically}. \tag{5.13}
$$

The commutation relations of the Cartesian components of orbital angular momentum $\vec{L}$ can also be readily derived as follows:

$$
\begin{aligned}
[L_x, L_y] &= [yp_z - zp_y, zp_x - xp_z] \\
&= [yp_z, zp_x] - [yp_z, xp_z] - [zp_y, zp_x] + [zp_y, xp_z]
\end{aligned} \tag{5.14}
$$

where the required commutators [A B C D] are derived as follows:

$$
\begin{aligned}
[AB, CD] &= [X, CD], \quad \text{where } X = AB \quad \text{(say)} \\
&= XCD - CDX \\
&= XCD - CXD + CXD - CDX \\
&= [X, C]D + C[X, D]
\end{aligned} \tag{5.15}
$$

$$
\begin{aligned}
[AB, Y] &= ABY - YAB \\
&= ABY - AYB + AYB - YAB \\
&= A[B, Y] + [A, Y]B.
\end{aligned} \tag{5.16}
$$

Therefore,

$$[AB, CD] = (A[B, C] + [A, C]B)D + C(A[B, D] + [A, D]B) \\ = A[B, C]D + [A, C]BD + CA[B, D] + C[A, D]B. \tag{5.17}$$

So,

$$[yp_z, zp_x] = y[p_z, z]p_x = -iyp_x \\ [yp_z, xp_z] = 0 = [zp_y, zp_x] \\ [zp_y, xp_z] = x[z, p_z]p_y = i(xp_y - yp_x) = iL_z. \tag{5.18}$$

Similarly,

$$[L_y, L_z] = iL_x \tag{5.19}$$

$$[L_z, L_x] = iL_y. \tag{5.20}$$

The three commutation relations for the components of $\vec{L}$ can be written in terms of the Levi-Civita symbol as:

$$[L_k, L_\ell] = i\varepsilon_{k\ell m}L_m, \tag{5.21}$$

where

$$\varepsilon_{k\ell m} = \begin{cases} +1, \text{ for even permutations of}(123) \\ -1, \text{ for odd permutations of}(123) \\ 0, \text{ otherwise.} \end{cases} \tag{5.22}$$

The square of the angular momentum operator:

$$L^2 = L_x^2 + L_y^2 + L_z^2 \tag{5.23}$$

and it commutes wit al the components of $\vec{L}$, so that

$$[L^2, L_k] = 0, \quad k = (123). \tag{5.24}$$

Quantum states can be specified by the simultaneous eigenfunctions of L^2 and L_z (or any one component of $(\vec{L})$). These (L^2, L_z) constitute the *complete set of commuting generators* for orbital angular momentum. If another component of $\vec{L}$ is included in this set, it will not commute with L_z. The measurement of another variable corresponding to an operator not commuting with the set L^2L_z, necessarily introduces uncertainty into one of the variables already measured. A sharper specification of the system is therefore not possible.

A general angular momentum operator $\vec{J}$ is defined as one whose Cartesian components obey the commutation relations:

$$[J_k, J_\ell] = i\varepsilon_{k\ell m}\, J_m. \tag{5.25}$$

This extended definition permits the existence of spin—a quantity that has no classical analogue.

In the case of orbital angular momentum $\vec{L}$, it is well known from the study of partial differential equations that the solution of the Laplace's equation,

$$\nabla^2\psi = 0, \tag{5.26}$$

is in spherical polar coordinates. A solution via the method of the separation of variables can be written as:

$$\psi(r, \theta, \phi) = R(r) Y_m^\ell(\theta, \phi) \tag{5.27}$$

where $Y_m^\ell(\theta, \phi)$ satisfies the differential equation:

$$\left[\frac{1}{\sin\theta}\frac{d}{d\theta}\left(\sin\theta\frac{d}{d\theta}\right) + \ell(\ell+1) - \frac{m^2}{\sin^2\theta}\right]\Theta_{\ell,m}(\theta) = 0 \tag{5.28}$$

and

$$\left[\frac{d^2}{d\phi^2} + m^2\right]\Phi_m(\phi) = 0 \tag{5.29}$$

with

$$Y_m^\ell(\theta, \phi) = \Theta_{\ell,m}(\theta)\Phi_m(\phi), \tag{5.30}$$

and are nothing but spherical harmonics, introduced previously.

In this representation, in analogy with the eigenvalue problem for orbital angular momentum, viz:

$$L^2 Y_m^\ell(\theta, \phi) = \ell(\ell+1)\hbar^2 Y_m^\ell(\theta, \phi) \tag{5.31}$$

$$L_z Y_m^\ell(\theta, \phi) = m\hbar Y_m^\ell(\theta, \phi), \tag{5.32}$$

which are also written in the Dirac bra-ket notation, in which the spherical harmonics satisfy the eigenvalue equations:

$$Y_m^\ell(\theta, \phi) = \Theta_{\ell,m}(\theta)\Phi(\phi) \equiv |\ell, m\rangle \tag{5.33}$$

$$L^2|\ell, m\rangle = \ell(\ell+1)|\ell, m\rangle \tag{5.34}$$

$$L_z|\ell, m\rangle = m|\ell, m\rangle. \tag{5.35}$$

For general angular momentum—which can be either orbital ($\vec{L}$), spin ($\vec{S}$), or their sum $\vec{J} = \vec{L} + \vec{S}$, called total angular momentum—one constructs the eigenstates of general angular momentum, in the bra-ket notation a the simultaneous eigenfunctions of J^2, J_z, as:

$$J^2|j, m\rangle = j(j+1)|j, m\rangle \tag{5.36}$$

$$J_z|j, m\rangle = |j, m\rangle \tag{5.37}$$

in the natural units $\hbar = c = 1$.

Note: in these equations, m stands for the projection of the angular momentum $\vec{J}$ and not mass (as in $E = mc^2$).

The operator

$$J_x^2 + J_y^2 + J_z^2 - J_z^2 \tag{5.38}$$

is diagonal in $|j, m\rangle$ representation and it has positive-definite (non-negative) eigenvalue:

$$(J_x^2 + J_y^2)|j, m\rangle = (J^2 - J_z^2)|j, m\rangle = (\lambda_j - m^2)|j, m\rangle, \tag{5.39}$$

because the expectation value of the square of a Hermitian operator, i.e. the square of a real eigenvalue, is greater than or equal to zero. Therefore

$$\lambda_j = j(j+1)\hbar^2 = \langle j, m|J^2|j, m\rangle, \quad \text{and} \quad \lambda_j \geqslant m^2. \tag{5.40}$$

Thus, the value of m is bounded by both above and below, and m^2 cannot exceed λ_j. This implies that, for a given $\vec{J}$, there exist minimum and maximum values of m, denoted by m_{min} and m_{max}.

Let $J_\pm$ define the raising and lowering operators:

$$J_\pm = J_x \pm iJ_y, \tag{5.41}$$

which satisfy the commutation relations:

$$[J^2, J_\pm] = 0 = [J^2, J_z] \tag{5.42}$$

and

$$[J_z, J_\pm] = \pm J_\pm \tag{5.43}$$

$$[J_+, J_-] = 2J_z. \tag{5.44}$$

Now examine the behaviour of the function $J_\pm|j, m\rangle$:

$$J^2 J_\pm|j, m\rangle = J_\pm J^2|j, m\rangle = \lambda_j J_\pm|j, m\rangle \tag{5.45}$$

$$J_z J_\pm|j, m\rangle = (J_\pm J_z \pm J_\pm)|j, m\rangle = (m \pm 1)J_\pm|j, m\rangle. \tag{5.46}$$

From these it follows that $J_{\pm}|j, m\rangle$ is an eigenfunction of J^2 with an eigenvalue λ_j, and also an eigenfunction of J_z with the eigenvalue $(m \pm 1)$.

Thus, $J_{\pm}|j, m\rangle$ is proportional to the normalized eigenfunction $|j, m \pm 1\rangle$, i.e.

$$J_{\pm}|j, m\rangle = C_{\pm}|j, m \pm 1\rangle, \tag{5.47}$$

where $C_{\pm}$ is a proportionality constant. The ability of the operators $J_{\pm}$ to raise/lower the value of m by the ± 1 unit, respectively, while preserving λ_j, gives them their nomenclature as raising/lowering, step-up/step-down, ladder or shift operators.

Since the values of m are bounded between $m_{\min}$ and $m_{\max}$, it follows that

$$J_{+}|j, m_{\max}\rangle = 0 = J_{-}|j, m_{\min}\rangle. \tag{5.48}$$

By applying J_- to the first and J_+ to the second and using:

$$J^2 = J_x^2 + J_y^2 + J_z^2 = \frac{1}{2}(J_+J_- + J_-J_+) + J_z^2 \tag{5.49}$$

$$= \frac{1}{2}(J_+J_- + J_+J_- - 2J_z) + J_z^2, \quad \text{since } [J_+, J_-] = 2J_z \tag{5.50}$$

$$J^2 = J_+J_- + J_z(J_z - 1), \tag{5.51}$$

so that

$$J_{\mp}J_{\pm} = J^2 - J_z(J_z + 1), \tag{5.52}$$

we obtain

$$J_-J_+|j, m_{\max}\rangle = (J^2 - J_z(J_z + 1))|j, m_{\max}\rangle \tag{5.53}$$

$$= \lambda_j - m_{\max}(m_{\max} + 1) = 0, \tag{5.54}$$

and

$$J_+J_-|j, m_{\min}\rangle = (J^2 - J_z(J_z - 1))|j, m_{\min}\rangle \tag{5.55}$$

$$= \lambda_j - m_{\min}(m_{\min} - 1) = 0. \tag{5.56}$$

Eliminating λ_j from these two equations:

$$m_{\max}(m_{\max} + 1) = m_{\min}(m_{\min} - 1) \tag{5.57}$$

or

$$(m_{\max} + m_{\min})(m_{\max} - m_{\min} + 1) = 0, \tag{5.58}$$

and one of the two factors must vanish. Since, $m_{\max} \geqslant m_{\min}$, the only possible solution is:

$$m_{\max} = -m_{\min}. \tag{5.59}$$

From $J_{\pm}|j, m\rangle = C_{\pm}|j, m \pm 1\rangle$, we also know that successive eigenvalues of m differ by unity. Therefore, $m_{\max} - m_{\min}$ is a positive definite integer, which we denote by 2j, where j is an *integer* or *half-integer* i.e.

$$m_{\max} - m_{\min} = 2j \qquad \text{and} \qquad m_{\max} + m_{\min} = 0, \tag{5.60}$$

so that $m_{\max}= j$, and $m_{\min}= -$j. Or,

$$-j \leqslant m \leqslant j, \tag{5.61}$$

$$i.\ e. \quad m = -j, -j+1, -j+2, \ldots, -1, 0, 1, 2, \ldots, j-2, j-1, j. \tag{5.62a}$$

Thus,

$$\lambda_j = m_{\max}(m_{\max}+1) = m_{\min}(m_{\min}-1) \tag{5.62b}$$

$$j(j+1) = -j(-j-1). \tag{5.63}$$

We are now in a position to evaluate the proportionality constant, $C_{\pm}$:

$$\langle j, m|J_{\mp} J_{\pm}| j, m\rangle = \langle j, m|J^2 - J_z(J_z \pm 1)| j, m\rangle \tag{5.64}$$

$$= j(j+1) - m(m \pm 1). \tag{5.65}$$

Note: in the theory of angular momentum, this replacement $j \to -j-1$ is a transformation, also referred to as Yutsis 'mirror' symmetry, which reflects the mathematical invariance of the eigenvalue $j(j+1)$ of the Casimir[11] operator J^2 for angular momentum. Though unphysical, this type of mirror symmetry have been studied by Yutsis and Bandzaitis (1977).

Furthermore,

$$\langle j, m|J_{\mp} J_{\pm}| j, m\rangle = \langle j, m|J_{\mp} \sum_{j',m'} | j', m'\rangle\langle j', m'|J_{\pm}| j, m\rangle, \tag{5.66}$$

using the completeness property:

$$\sum_{\nu} |\nu\rangle\langle\nu| = 1, \tag{5.67}$$

[11] Hendrik Brugt G Casimir (1909–2000), Dutch physicist.

$$
\begin{aligned}
\langle j, m|J_{\mp} J_{\pm}|J_{\pm}|j, m\rangle &= \langle j, m|J_{\mp}|j, m \pm 1\rangle\langle j, m \pm 1|J_{\pm}|j, m\rangle \\
&= (\langle j, m \pm 1|J_{\pm}|j, m\rangle)^{\dagger}\langle j, m \pm 1|J_{\pm}|j, m\rangle \\
&= C_{\pm}^{\dagger}C_{\pm} = |C_{\pm}|^2.
\end{aligned}
\tag{5.68}
$$

The absolute value of $C_{\pm}$ is determined up to an arbitrary phase. The Condon and Shortley convention adopts the positive root. So,

$$C_{\pm} = [j(j+1) - m(m \pm 1)]^{1/2} = [(j \mp 1)(j \pm m + 1)]^{1/2}, \tag{5.69}$$

i.e. $C_{\pm}$ is real and the matrix elements of J_x are real while those of J_y are purely imaginary:

$$\langle j', m'|J^2|j, m\rangle = j(j+1)\delta_{jj'}\delta_{mm'} \tag{5.70}$$

$$\langle j', m'|J_z|j, m\rangle = m\delta_{jj'}\delta_{mm'} \tag{5.71}$$

$$\langle j', m'|J_{\pm}|j, m\rangle = [(j \mp m)(j \pm m + 1)]^{1/2}\delta_{j'j}\delta_{m',m\pm 1}. \tag{5.72}$$

The last of the above can be written also as:

$$\langle j', m'|J_x|j, m\rangle = \frac{1}{2}[(j \mp m)(j \pm m + 1)]^{1/2}\delta_{j'j}\delta_{m',m\pm 1} \tag{5.74}$$

$$\langle j', m'|J_y|j, m\rangle = \mp\frac{i}{2}[(j \mp m)(j \pm m + 1)]^{1/2}\delta_{j'j}\delta_{m',m\pm 1} \tag{5.75}$$

where $\delta_{m,\,n}$ is the Kronecker delta function:

$$\delta_{m,\,n} = \begin{cases} 0 & m \neq n \\ 1 & m = n. \end{cases} \tag{5.76}$$

5.3 Representations of angular momentum operators

As an exercise, let us find the two-dimensional (and three-dimensional) representations of the angular momentum operators:

$\langle m'|J_z|m\rangle$: (5.77)

m'/m	$+\frac{1}{2}$	$-\frac{1}{2}$
$+\frac{1}{2}$	$\frac{1}{2}$	0
$-\frac{1}{2}$	0	$-\frac{1}{2}$

$\langle m'|J_+|m\rangle$: (5.78)

$m' = m + 1/m$	$+\frac{1}{2}$	$-\frac{1}{2}$
$+\frac{1}{2}$	0	1
$-\frac{1}{2}$	0	0

and

$\langle m'|J_-|m\rangle$: (5.79)

$m' = m-1/m$	$+\frac{1}{2}$	$-\frac{1}{2}$
$+\frac{1}{2}$	0	0
$-\frac{1}{2}$	1	0.

Therefore

$$J_z = 1/2\begin{pmatrix}1 & 0\\ 0 & -1\end{pmatrix};\quad J_+ = 1/2\begin{pmatrix}0 & 1\\ 0 & 0\end{pmatrix};\quad J_- = 1/2\begin{pmatrix}0 & 0\\ 1 & 0\end{pmatrix}; \tag{5.80}$$

since $J_\pm = J_x \pm iJ_y$

$$J_x = \frac{1}{2}(J_+ + J_-) = \frac{1}{2}\begin{pmatrix}0 & 1\\ 1 & 0\end{pmatrix} \tag{5.81}$$

$$J_y = \frac{1}{2i}(J_+ - J_-) = \frac{1}{2i}\begin{pmatrix}0 & 1\\ -1 & 0\end{pmatrix} = \frac{1}{2}\begin{pmatrix}0 & -i\\ i & 0\end{pmatrix}. \tag{5.82}$$

Note: when $J_k = \frac{1}{2}\sigma_k$

$$\sigma_z = \begin{pmatrix}1 & 0\\ 0 & -1\end{pmatrix},\quad \sigma_x = \begin{pmatrix}0 & 1\\ 1 & 0\end{pmatrix},\quad \sigma_y = \begin{pmatrix}0 & -i\\ i & 0\end{pmatrix} \tag{5.83}$$

the σ_k, $(k = 1,2,3)$ matrices are the Pauli spin-1/2 matrices.

The corresponding spin 1 matrices are:

$\langle m'|J_z|m\rangle$: (5.84)

m'/m	$+1$	0	-1
$+1$	1	0	0
0	0	0	0
-1	0	0	-1

$\langle m-1|J_+|m\rangle$: (5.85)

$m-1/m$	$+1$	0	-1
$+1$	0	1	0
0	0	0	1
-1	0	0	0

and

$\langle m'|J_-|m\rangle$: (5.86)

$m'/m-1$	$+1$	-1	
$+1$	0	0	0.
0	0	0	1
-1	0	0	0

Therefore,

$$J_z = \begin{pmatrix} 1 & 0 & 0 \\ 0 & 0 & 0 \\ 0 & 0 & -1 \end{pmatrix}; \quad J_+ = \begin{pmatrix} 0 & 1 & 0 \\ 0 & 0 & 0 \\ 0 & 0 & 0 \end{pmatrix}; \quad J_- = \begin{pmatrix} 0 & 0 & 0 \\ 1 & 0 & 0 \\ 0 & 1 & 0 \end{pmatrix}; \tag{5.87}$$

since $J_\pm = J_x \pm iJ_y$

$$J_x = \frac{1}{2}(J_+ + J_-) = \frac{1}{2}\begin{pmatrix} 0 & 1 & 0 \\ 1 & 0 & 1 \\ 0 & 1 & 0 \end{pmatrix} \tag{5.88}$$

$$J_y = \frac{1}{2i}(J_+ - J_-) = \frac{1}{2}\begin{pmatrix} 0 & -i & 0 \\ i & 0 & -i \\ 0 & i & 0 \end{pmatrix}. \tag{5.89}$$

The operator $J^2 = J_x^2 + J_y^2 + J_z^2$ is a *Casimir operator*, which commutes with all other generators: $[J^2, J_i] = 0$. Also,

$$J_z\, J_\pm | j, m\rangle = (m \pm 1) J_z | j, m\rangle \tag{5.90}$$

and

$$J_\pm | j, m\rangle = \sqrt{(j \mp m)(j \pm m + 1} \,| j, m \pm 1\rangle. \tag{5.91}$$

The angular momentum commutation relations:

$$[J_a, J_b] = i\varepsilon_{abc}\, J_c \tag{5.92}$$

represent the simplest *non-Abelian Lie algebra*, with ε_{abc} being the *structure constants*. The summation convention implies that there is a sum over c, but, in this case, it is trivial due to the properties of the Levi-Civita tensor and the commutator $[J_1, J_2]$ giving the 'other' (third) generator J_3, e.g. $[J_1, J_2] = i\, J_3$, and it is *Lie algebra*. Note: $J_i = \frac{1}{2}\sigma_i$, $i = 1, 2, 3$, the generators, provide a suitable (but not unique) representation of the Pauli matrices σ_i, which have the properties:

$$\sigma_i \sigma_j = \delta_{ij} + i\varepsilon_{ijk}\, J_k. \tag{5.93}$$

Irreducible representations are characterized by the highest $J_z \equiv J_3$ eigenvalue, j, which must be an integer or half-odd integer, and they are (2j + 1)-dimensional.

The eigenvalues of J_3 are called *weights* and j is the highest weight.

The construction for angular momentum, characterizing the irreducible representations (irreps or IRs), in terms of their highest weight, generalizes to any Lie algebra. The representation with the highest weight j is sometimes called the *spin j representation.*

When Hermitian angular momentum operators exist, the entire Hilbert space is a representation and can be decomposed into states labeled by $|\alpha\, j\, m\rangle$ in the spin j representation with $J_z = m$. The α label stands for all other labels necessary to characterize the states. The state can be chosen to be orthonormal in α:

$$\langle \alpha' j' m' | \alpha j m\rangle = \delta_{\alpha,\,\beta}\delta_{j,\,j'}\delta_{m,\,m'}. \tag{5.94}$$

These states form a *basis* for studying the properties of the system under rotations.

Just as any representation can be broken down into simpler irreducible representations, an arbitrary representation can be built out of the simplest representation. The simplest non-trivial representation is generated by the Pauli matrices:

$$J_a = \sigma_a/2; \quad \sigma_1 = \begin{pmatrix} 0 & 1 \\ 1 & 0 \end{pmatrix}; \quad \sigma_2 = \begin{pmatrix} 0 & -i \\ i & 0 \end{pmatrix}; \quad \sigma_3 = \begin{pmatrix} 1 & 0 \\ 0 & -1 \end{pmatrix}. \tag{5.95}$$

The group elements, $\exp[i\lambda_a \sigma_a/2]$ are the special unitary 2×2 matrices. The group is called $SU(2)$, which stands for special unitary unimodular group.

In mathematics, the special unitary group of degree n, denoted by $SU(n)$, is the Lie group of $n \times n$ unitary matrices with the determinant 1. The group operation is matrix multiplication. The special unitary group is a subgroup of the unitary group

$U(n)$, consisting of all $n \times n$ unitary matrices. As a compact classical group, $U(n)$ is the group that preserves the standard inner product on C^n. It is itself a subgroup of the general linear group, $SU(n) \subset U(n) \subset GL(n, C)$.

The $SU(n)$ groups find wide application in the standard model of particle physics, especially $SU(2)$ in the electroweak interaction and $SU(3)$ in quantum chromo dynamics (see Arima (1999)).

The simplest case, $SU(1)$, is the trivial group, which has only a single element. The group $SU(2)$ is isomorphic to the group of quaternions of norm 1, and is thus diffeomorphic to the 3-sphere. Since unit quaternions can be used to represent rotations in three-dimensional space (up to sign), there is a surjective homomorphism from $SU(2)$ to the rotation group $SO(3)$. $SU(2)$ is also identical to one of the symmetry groups of spinors, *Spin*(3), that enables a spinor presentation of rotations.

Given the two representations, $D_1(n)$ and $D_2(m)$, of a group, it is possible to form another representation of the dimension $m + n$ as their direct sum, which is formed by the block diagonal matrices

$$\begin{bmatrix} D_1(g) & 0 \\ 0 & D_2(g) \end{bmatrix}. \tag{5.96}$$

One can also form a representation of the dimension $n \times m$ as follows. If $|i\rangle$, $i = 1, 2, \ldots, n$ is an orthonormal basis on which D_1 acts and $|x\rangle$, $x = 1, 2, \ldots, n$ is an orthonormal basis on which D_2 acts, then the product $|i\rangle|x\rangle$ is an orthonormal basis in an $n \times m$ dimensional space called the *direct product* space. On this space, the direct product is:

$$(D_1 \otimes D_2)(g)\{|i\rangle|x\rangle\} = D_1(g)|i\rangle D_2(g)|x\rangle. \tag{5.97}$$

In matrix language:

$$\langle y|\langle j|(D_1 \otimes D_2)(g)|i\rangle|x\rangle = \langle j|D_1(g)|i\rangle\langle j|D_2(g)|x\rangle, \tag{5.98}$$

$$i.\ e.\ [(D_1(g) \otimes D_2(g)]_{ij,\ xy} = [D_1(g)]_{ij}[D_2(g)]_{xy}. \tag{5.99}$$

We can form a direct product of n two-dimensional representations as

$$[D \otimes \cdots \otimes D]_{i_1 i_2 \cdots i_n,\ j_1 \cdots j_n} = D_{i_1 j_1} \cdots D_{i_n j_n} \tag{5.100}$$

acting on the n-component objects $U_{j_1 \cdots j_n}$. This representation is reducible.

5.4 The rotation group

Consider $\mathcal{R}^3$, the real three-dimensional vector space:

$$\mathcal{R}^3 = \{\vec{x} = (x_1, x_2, x_3 \mid x_i \text{real})\}.$$

The linear transformations:

$$\vec{x} \to R\vec{x} \qquad \text{or} \qquad x' = R_{ij}x_j, \quad i, j = 1, 2, 3 \tag{5.101}$$

which preserve the scalar product:

$$(x', y') = \sum_i x_i y_i \tag{5.103}$$

are such that:

$$(x', y') = R_{ij} x_j R_{ik} y_k = R_{ij} R_{ik} x_j y_k \equiv (x, y), \tag{5.104}$$

which implies

$$\sum_{j,k} R_{ij} R_{ik} = \delta_{j,k} \qquad \text{or} \qquad RR^T = I, \tag{5.105}$$

where I is the 3×3 unit matrix. Hence, R is real and orthogonal.

The set of all real, orthogonal matrices with $\det R = +1$ form the subgroup $SO(3)$ of the special orthogonal transformations in real three-dimensional space. They do not include inversions (or space reflections). The set of all matrices with $\det R = -1$ do not form a subgorup, since the identity does not belong to this subset, which is not closed under multiplication.

Let the element

$$P = \begin{pmatrix} -1 & 0 & 0 \\ 0 & -1 & 0 \\ 0 & 0 & -1 \end{pmatrix} \varepsilon O(3) \tag{5.106}$$

be the parity operator with

$$\det P = -1 \qquad \text{and} \qquad P^2 = I. \tag{5.107}$$

If $R \;\varepsilon\; O(3)$ and $\det R = -1$, let

$$R = PR', \qquad \text{where} \qquad R' = PR. \tag{5.108}$$

Now $R' \;\varepsilon\; O(3)$ and $\det R' = \det(PR) = +1$, which implies that $R' \;\varepsilon\; SO(3)$. Therefore,

$$O(3) = SO(3) \cup PSO(3). \tag{5.109}$$

The corresponding groups in the n-dimensions are $O(n)$ and $SO(n)$, with

$$O(n) = SO(n) \cup PSO(n) \tag{5.110}$$

where $P = -$ I, I being the 3×3 unit matrix.

The group SU(2) consists of 2 x 2 unitary, unimodular matrices in two-dimensional complex vector space, which leaves

$$(z, z') = \sum_i z_i^* z'_i \quad \text{invariant.} \tag{5.111}$$

A similar definition holds for $SU(n)$, where $g \varepsilon SU(n)$ is a $n \times n$ unitary, unimodular matrix.

5.5 The ${}_3F_2(1)$ sets

In quantum mechanics, consider the addition of two angular momenta $\overrightarrow{j_1}$, $\overrightarrow{j_2}$ to get a total angular momentum $\overrightarrow{j_3}$:

$$\overrightarrow{j_1} + \overrightarrow{j_2} = \overrightarrow{j_3}. \tag{5.112}$$

For example, an electron in the Bohr atom model: when it is in an orbit of angular momentum $\overrightarrow{L}$ and has an intrinsic spin $\overrightarrow{S}$, its total angular momentum is an electron in the Bohr atom model; but when it is in an orbit of angular momentum $\overrightarrow{L}$ and has an intrinsic spin of $\overrightarrow{S}$, the total angular momentum is:

$$\overrightarrow{J} = \overrightarrow{L} + \overrightarrow{S}. \tag{5.113}$$

In the 'classical' theory of angular momentum, $\overrightarrow{L}$ takes integral values, $\overrightarrow{S}$ is half-integral, and, consequently, $\overrightarrow{J}$ can take only half-integral values. The numerical values of $\overrightarrow{J}$ are:

$$|L - S| \leqslant J \leqslant L + S. \tag{5.114}$$

(5.114) is called the triangle inequality and it is equivalent to:

$$L + S - J \geqslant 0, \quad L - S + J \geqslant 0, \quad L + S - J \geqslant 0. \tag{5.115}$$

The uncoupled system can be represented by the complete set of commuting generators:

$$L^2,\ S^2,\ L_z,\ S_z \tag{5.116}$$

and the coupled system by the complete set of commuting generators

$$L^2,\ S^2,\ J^2,\ J^z \tag{5.117}$$

where L^2, S^2, and J^2 are the Casimir operators, which commute with their corresponding Cartesian components, viz

$$[L^2, L_k] = 0, \quad [S^2, S_k] = 0, \quad [J^2, J_k] = 0. \tag{5.118}$$

Let us denote the orthonormal basis vector for the uncoupled system in the Dirac bra-ket notation by $|L, \mu\rangle$, $|S, \nu\rangle$ and the orthonormal basis for the coupled state by $|LSJM\rangle$. These two orthonormal basis vectors are related to each other by an orthogonal transformation

$$|LSJM\rangle = \sum_{\mu,\nu} C(LSJ;\ \mu\nu M)|L\mu\rangle|S\nu\rangle, \tag{5.119}$$

where $C(L\ S\ J;\ \mu\ \nu\ M)$ is the transformation coefficient, which became known as the Clebsch–Gordan coefficient after the work of Clebsch[12] and Gordan[13] on the

[12] Dudolf Friedrich Alfred Clebsch (1833–1872), German mathematician.
[13] Paul Albert Gordan (1837–1912), German mathematician.

invariant theory of algebraic forms, which is an equivalent formulation of the coupling of two angular momenta. In physics literature, these are also synonymously referred to as the *vector addition* or *vector coupling* coefficients. These were designated as Wigner coefficients by Biedenharn and Louck (1981).

The Clebsch–Gordan coefficient is non-vanishing only when the triangle inequality (5.114) is satisfied and the projection quantum numbers μ, ν, M obey the additive law

$$M = \mu + \nu. \tag{5.120}$$

Quantum theory of angular momentum has established the existence of an inverse of (5.119) as

$$|L\mu\rangle|S\nu\rangle = \sum_{J,M} C(LSJ;\, \mu\nu M)|LSJM\rangle \tag{5.121}$$

obtained using the orthogonality properties satisfied by the Clebsch–Gordan coefficients

$$\sum_{\mu,\nu} C(LSJ;\, \mu\nu M)C(LSJ';\, \mu\nu M') = \delta_{JJ'}\delta_{JJ'}\delta_{MM'}. \tag{5.122}$$

In 1940, Eugene Paul Wigner defined the 3-*j* symbol, or 3-*j* coefficient, of angular momentum as

$$\begin{pmatrix} j_1 & j_2 & j_3 \\ m_1 & m_2 & m_3 \end{pmatrix} = \frac{(-1)^{j_1-j_2-m_3}}{[j_3]} C(j_1 j_2 j_3;\, m_1 m_2 - m_3), \tag{5.123}$$

where $[j_3] = \sqrt{2j_3+1}$ and the projection quantum numbers in the 3-*j* coefficient satisfy the condition:

$$m_1 + m_2 + m_3 = 0. \tag{5.124}$$

The Wigner 3-*j* coefficient is defined as

$$\begin{aligned} \begin{pmatrix} j_1 & j_2 & j_3 \\ m_1 & m_2 & m_3 \end{pmatrix} &= \delta_{m_1+m_2+m_3,\, 0}(-1)^{j_1-j_2-m_3}\Delta(j_1 j_2 j_3) \\ &\times \prod_{i=1}^{3} \left[(j_i + m_i)!(j_i - m_i)!\right]^{1/2} \\ &\times \sum_t (-1)^t \left[t! \prod_{k=1}^{2} (t - \alpha_k)! \prod_{\ell=1}^{3} (\beta_\ell - t)!\right]^{-1} \end{aligned} \tag{5.125}$$

where

$$t_{\min} \leqslant t \leqslant t_{\max} \tag{5.126}$$

$$t_{\min} = \max(0, \alpha_1, \alpha_2), \qquad t_{\max} = \min(\beta_1, \beta_2, \beta_3) \tag{5.127}$$

$$\begin{aligned} \alpha_1 &= j_1 - j_3 + m_2 = (j_1 - m_1) - (j_3 + m_3) \\ \alpha_2 &= j_2 - j_3 - m_1 = (j_2 + m_2) - (j_3 - m_3) \end{aligned} \tag{5.128}$$

$$\beta_1 = j_1 - m_1, \quad \beta_2 = j_2 + m_2, \quad \beta_3 = j_1 + j_2 - j_3$$

and

$$\Delta(xyz) = \left[\frac{(-x+y+z)!(x-y+z)!(x+y-z)!}{(x+y+z+1)!}\right]^{\frac{1}{2}}. \tag{5.130}$$

The function $\Delta(xyz)$ vanishes unless the usual triangle inequality (5.114) is satisfied by the three angular momenta x, y, z.

5.6 Symmetries of the 3-*j* coefficient

For several notations of the Clebsch–Gordan and other angular momentum coefficients, refer to p 150 of Biedenharn and Louck (1981), wherein several derivations of the Clebsch–Gordan coefficient are also discussed.

The series part of the Clebsch–Gordan, or the 3-*j* coefficient, in (5.125) clearly exhibits 12 symmetries, since it is invariant under the permutations of the two α-parameters and the three β-parameters (or $2! \times 3! = 12$). However, it should be noted that these are not the so called 'classical' symmetries of the 3-*j* coefficient, which were known to exist from the very beginning due to the invariance of the 3-*j* coefficient to its three permutations of j_1, j_2, j_3 and the space reflection

$$m_i \to -m_i, \quad i = 1, 2, 3. \tag{5.131}$$

In the 1950s, when tables of angular momentum coefficients were widely referred to, symmetries of these coefficients were useful to curtail their sizes. The tables of 3-*j* (and 6-*j*) symbols were provided by Rotenberg *et al* (1959). A Wigner 3-*j*, 6-*j*, and 9-*j* calculator can be found on the internet, which calculates these symbols exactly, and has been developed by Anthony Stone. This is available at: www-stone.ch.cam.ac.uk/wigner.shtml[14].

In 1958, Tulio Regge[15] made a dramatic discovery of new symmetry properties for the 3-*j* coefficient. He arranged the nine non-negative integer parameters, referred to by Racah (1942a, 1942b), which can be formed from the three J_is and M_is, viz:

$$j_1 - m_1, \quad j_2 - m_2, \quad j_3 - m_3, \quad j_1 + m_1, \quad j_2 + m_2, \quad j_3 + m_3, \tag{5.132}$$

into a 3×3 square symbol and reOresented the 3-*j* coefficient as (Regge 1958):

[14] See also Johansson and Forssen (2015).

[15] Tulio Eugenio Regge (1931–2014), Italian physicist.

$$\begin{pmatrix} j_1 & j_2 & j_3 \\ m_1 & m_2 & m_3 \end{pmatrix} = \left\| \begin{matrix} -j_1 + j_2 + j_3 & j_1 - j_2 + j_3 & j_1 + j_2 - j_3 \\ j_1 - m_1 & j_2 - m_2 & j_3 - m_3 \\ j_1 + m_1 & j_2 + m_2 & j_3 + m_3 \end{matrix} \right\| \equiv |R_{ik}|| \tag{5.133}$$

and noted that all sums of the columns and rows add to $J = j_1 + j_2 + j_3$, which is a property of a class of magic squares. Regge asserted that the 3-j coefficient has 72 symmetries, being invariant to the three column permutations, three row permutations, and to a reflection about the diagonality of the 3×3 Regge square symbol. Of these, the well-known *classical symmetries* arise due to the three row permutations and to the exchange of rows 2 and 3 in the $||R_{ik}||$. Regge stated explicitly that 'we cannot justify these symmetries using physical arguments'. He also did not write down the six new symmetries explicitly.

Racah has shown that, assuming the argument of one of the five factorials in (5.125) are the summation index instead of ℓ, leads to some symmetry properties of the Clebsch–Gordan coefficient. We have shown (Srinivasa Rao 1978) that making such a substitution successively for each of the five factorials in (5.125) results in five series representations. These five, along with the original (5.125)—which alone is given in literature conventionally till 1978—constitute a set of six series representations, which can also be obtained by permuting the indices (123) of the j_is and m_is in (5.125). Since the 72-element symmetry group is evident when the 3-j coefficient is represented by the 3×3 Regge symbol $||R_{ik}||$, we define the set of six series representations in terms of the $||R_{ik}||$s as

$$\begin{aligned}\begin{pmatrix} j_1 & j_2 & j_3 \\ m_1 & m_2 & m_3 \end{pmatrix} &= \delta_{m_1+m_2+m_3,\,0}(-1)^{\sigma(pqr)} \prod_{i,k=1}^{3} \frac{R_{ik}!}{(J+1)!} \\ &\times \sum_s (-1)^s \big[s!(R_{2p}-s)!(R_{3q}-s)!(R_{1r}-s)! \\ &\times (s+R_{3r}-R_{2p})!(s+R_{2r}-R_{3q})!\big]^{-1}\end{aligned} \tag{5.134}$$

for all six permutations of $(pqr) = (123)$ with

$$\sigma(pqr) = \begin{cases} R_{3p} - R_{2q} & \text{for even permutation of (123)} \\ R_{3p} - R_{2q} + J & \text{for odd permutation of (123).} \end{cases} \tag{5.135}$$

The six column permutations are in one-to-one correspondence with the six series representations, thereby spanning the whole set given by (5.134). Each series representation exhibits 12 of the 72 distinctly different symmetries of the 3-j coefficient. This 12-element symmetry group is isomorphic to the three permutations of the three objects R_{2p}, R_{3q}, and R_{1r}, and the two permutations of the two objects $(R_{3r} - R_{2p}, R_{2r} - R_{3q})$. The same set of six series representations can also be obtained by permuting the indices (123) in the expansion for the 3-j coefficient given by (5.123) and (5.125), and remembering that the series acquires an additional phase factor of $(-1)^J$ for odd permutations.

Rose, in 1955, pointed out that the Clebsch–Gordan coefficient given by the series (5.125) can be expressed in terms of a ${}_3F_2(1)$ hypergeometric function of the unit argument. To this end, the factorials are replaced by gamma functions, since

$$n! = \Gamma(n + 1), \tag{5.136}$$

and whenever the summation index in the argument of the gamma function is negative, using

$$\Gamma(z)\Gamma(1 - z) = \pi \csc \pi z, \tag{5.137}$$

it is replaced by a gamma function containing a positive index of summation through

$$\Gamma(1 - z - n) = (-1)^n \frac{\Gamma(z)\Gamma(1 - z)}{\Gamma(z + n)}. \tag{5.138}$$

Thus, from (5.134) we obtain a set of six ${}_3F_2(1)$s as

$$\begin{pmatrix} j_1 & j_2 & j_3 \\ m_1 & m_2 & m_3 \end{pmatrix} = \delta_{m_1+m_2+m_3,\, 0}(-1)^{\sigma(pqr)} \prod_{i,k=1}^{3} \left[\frac{R_{ik}!}{(J + 1)!}\right]^{1/2} \times [\Gamma(1 - A, 1 - B, 1 - C, D, E)]^{-1} {}_3F_2(A, B, C; D, E; 1) \tag{5.139}$$

where

$$\Gamma(x, y, \ldots) = \Gamma(x)\Gamma(y)\cdots, \tag{5.140}$$

$$A = -R_{2p},\ B = -R_{3q},\ C = -R_{1r},\ D = 1 + R_{3r} - R_{2p},\ E = 1 + R_{2r} - R_{3q}, \tag{5.141}$$

for all permutations of $(pqr) = (123)$. It is to be noted that all the three numerator variables in the ${}_3F_2(1)$s in (5.139) are negative integers and hence the ${}_3F_2(1)$s are the terminating series with the number of terms in the series determined by $\min(|a|, |B|, |C|)$. In Smorodinskii and Shelepin (1972), (see also ... Biedenharn and Louck (1981), only one member of this set, corresponding to $(pqr) = (123)$ is given and this set, which is necessary and sufficient to account for the 72 symmetries of the 3-j coefficient, has been found by Srinivasa Rao (1978).

On the basis of group theoretical methods, Wigner, in his book on quantum physics and group and group theory (see Wigner 1959), derived the form for the 3-j coefficient as:

$${}_3F_2\left(\begin{matrix} j_1 - m_1+1, -j_3 - m_3, j_1 - j_2 - j_3; 1 \\ -j_2 - j_3 - m_1, j_1 - j_2 - m_3+1 \end{matrix}\right). \tag{5.142}$$

Racah (1942a, 1942b) gave an algebraic derivation of the 3-j coefficient using certain recurrence relations and his ${}_3F_2(1)$ form is

$$ {}_3F_2\left(\begin{matrix} j_1 + m_1+1, -j_3 + m_3, -j_1 + m_1; 1 \\ -j_2 - j_3 + m_1, j_2 - j_3 + m_1+1 \end{matrix}\right). \tag{5.143} $$

Majumdar (1958) defined new one-variable operators for angular momentum, using a set of first order differential equations, whose solutions involve hypergeometric functions and a general expression followed from the four known properties of these functions and it is:

$$ {}_3F_2\left(\begin{matrix} j_1 + j_2 - j_3+1, -j_3 - m_3, j_1 - j_2 - j_3; 1 \\ -2j_3, j_1 - j_3 - m_2+1 \end{matrix}\right). \tag{5.144} $$

We call these as the Wigner, Racah, and Majumdar ${}_3F_2(1)$ forms of the 3-j coefficient.

5.7 Inter-relationship between the sets of ${}_3F_2(1)$s

Referring to the then known forms of ${}_3F_2(1)$s for the 3-j coefficient, Biedenharn, and Louck (1984, chapter 3, p 76) stated that 'they can be transformed one into another by symmetry transformations and/or a transformation method introduced by Racah.' Here, we show how the transformation theory of generalized hypergeometric series has been used by Srinivasa Rao (1990) to derive the Wigner, Racah, and Majumdar ${}_3F_2(1)$ sets for the 3-j coefficient from the more symmetric van der Waerden set of six ${}_3F_2(1)$s. The Weber–Erdelyi (W–E) transformation formula can be proved simply from the integral representation for the ${}_3F_2(1)$

$$ {}_3F_2(a, b, -n; c, d; 1) = \frac{\Gamma(d)}{\Gamma(b, d-b)} \times \int_0^1 {}_2F_1(a, -n; d; t) t^{b-1}(1-t)^{d-b-1} dt. \tag{5.145} $$

We use the well-known hypergeometric function identity

$$ {}_2F_1(a, -n; c; t) = \frac{\Gamma(c, c-a+n)}{\Gamma(c-a, c+n)} {}_2F_1(a, -n; a-c-n+1; 1-t), \tag{5.146} $$

then substitute the variable t by 1−t and use the transformation (5.145) again to derive the ${}_3F_2(1)$ transformation formula

$$ {}_3F_2(a, -n; c, d; 1) = \frac{\Gamma(c, c-a+n)}{\Gamma(c+n, c-a)}{}^3 F_2(a, d-b, -n; 1+a-c-n, d; 1), \tag{5.147} $$

where n is an integer, which determines the number of terms in the ${}_3F_2(1)$ series. We will refer to (5.147) as the Weber–Erdelyi (W–E) transformation formula (which can be derived from tables II_A and II_B, in Bailey (1935) obtained by Thomae in the notation of Whipple) and it corresponds to

$$F_p(0;\,4,\,5) = (-1)^m \frac{\Gamma(\alpha_{124},\,\alpha_{014},\,\alpha_{024})}{\Gamma(\alpha_{123},\,\alpha_{124},\,\alpha_{125})} F_n(4;\,0,\,1). \tag{5.148}$$

Identifying the numerator and denominator variables of the van der Waerden ${}_3F_2(1)$ functions given in the transformation (5.139) as

$$a = A, \quad b = B, \quad n = -C, \quad c = D, \quad d = E \tag{5.149}$$

and applying the W–E transformation (5.144), we will get the 3-j coefficient

$$\begin{aligned}\begin{pmatrix} j_1 & j_2 & j_3 \\ m_1 & m_2 & m_3 \end{pmatrix} &= \delta_{m_1+m_2+m_3,\,0}\,(-1)^{\sigma(pqr)} \prod_{i,k=1}^{3} \left[\frac{R_{ik}!}{(J+1)!}\right]^{1/2} \\ &\times \Gamma(1-D')[\Gamma(1-A',\,1-C',\,1+B'-E')]^{-1} \\ &\times [\Gamma(E',\,1+A'-D',\,1+C'-D')]^{-1}{}_3 \\ &F_2(A',\,B',\,C';\,D',\,E';\,1),\end{aligned} \tag{5.150}$$

where

$$\begin{aligned} A' &= -R_{2p},\ B' = 1 + R_{2r},\ C' = -R_{1r},\ D' \\ &= -R_{1r} - R_{3r},\ E' = 1 + R_{2r} - R_{3q}. \end{aligned} \tag{5.151}$$

This set of ${}_3F_2(1)$ functions are the Wigner set of ${}_3F_2(1)$, since, in (5.151), setting $(pqr) = (132)$ results in the Wigner form of the 3-j coefficient given by Raynal (1978, eq.(28)).

Alternatively, identifying the variables in (5.139) as

$$a = B, \quad b = C, \quad n = -A, \quad c = D, \quad x = E \tag{5.152}$$

and making use of the W–E transformation (5.144), we will get the form (5.151) for the 3-j coefficient with

$$\begin{aligned} A' &= -R_{2p},\ B' = 1 + R_{3p},\ C' = -R_{3q},\ D' \\ &= -R_{3q} - R_{3r},\ E' = 1 + R_{2r} - R_{3q}. \end{aligned} \tag{5.153}$$

This set is the Racah set of ${}_3F_2(1)$ functions, since, in the (5.151) setting, $(pqr) = (132)$ gives the Racah form of the 3-j coefficient.

Finally, an identification for the parameters in (5.139) as

$$a = C, \quad b = A, \quad n = -B, \quad c = D, \quad x = E \tag{5.154}$$

and use of the W–E transformation (5.144) will yield the 3-j coefficient of the form (5.150) but with the numerator and denominator variables being

$$\begin{aligned} A' &= -R_{1r},\ B' = 1 + R_{1q},\ C' = -R_{3q},\ D' \\ &= -R_{2q} - R_{3q},\ E' = 1 + R_{2r} - R_{3q}. \end{aligned} \tag{5.155}$$

This set of functions are the Majumdar set, since for $pqr = (321)$, the Majumdar form of the 3-j coefficient is obtained.

Thus, we have shown that starting with the highly symmetric van der Waerden set of ${}_3F_2(1)$, three sets of ${}_3F_2(1)$, corresponding to the Wigner, Racah, and Majumdar forms, can be obtained by simply using the Weber–Erdelyi transformation with different identifications for the numerator variables. Conversely, the same Weber–Erdelyi transformation can be used to get the Van der Waerden set from the Wigner, Racah, and Majumdar sets, by virtue of the fact that the matrix relating the numerator and denominator variables of the ${}_3F_2(1)$s in (5.145) acts like a projection operator.

Corresponding to the three identifications made above for the numerator and denominator variables—viz (5.147), (5.150) and (5.152)—we can make three more identifications with

$$c = E, \qquad d = D. \tag{5.156}$$

That is, the interchange of the denominator variables can also be shown to result in the same three sets of ${}_3F_2(1)$s, but in a different order, viz the Majumdar, Racah, and Wigner forms given by (5.155), (5.153) and (5.151), respectively, on which are superposed:

(i) the interchange of the p, q indices, and
(ii) the $m_i \to -m_i$ mirror symmetry.

Also, starting with a given ${}_3F_2(1)$ belonging to the van der Waerden set, and resorting to the work of Whipple (1925) on the symmetries of the ${}_3F_2(1)$ functions, Raynal (1978) showed that the ${}_3F_2(1)$ forms of Wigner, Racah, and Majumdar can be obtained.

In the case of the van der Waerden set of ${}_3F_2(1)$s (5.139), all three numerator variables are negative integer variables, and the two denominator variables are positive integers. But, in the case of the Wigner (5.140), Racah (5.141), and Majumdar (5.142), for the sets of ${}_3F_2(1)$s, two of the three numerator variables, A' and C', are negative integer variables, while the third B' is a positive integer. Of the two denominator variables, D' is always a negative integer, and the other E' is always a positive integer variable—a property that follows from the triangle inequality and the identity satisfied by the elements of the 3×3 Regge square symbols $\|R_{ik}\|$:

$$R_{\ell p} + R_{mp} = R_{nq} + R_{nr}, \tag{5.157}$$

for cyclic permutations of both (ℓmn) and (pqr) equal to (123).

In the case of the van der Waerden set of six ${}_3F_2(1)$s, all three numerator and two denominator variable permutations are allowed, as is manifestly evident from (5.139). For each member of the set, these permutations account for 12 symmetries, and hence for the whole set of 72 symmetries, each of the 3-j coefficients will be accounted for. But, in the case of the Wigner, Racah, and Majumdar sets of six ${}_3F_2(1)$s, each member of the ${}_3F_2(1)$ set accounts for only two symmetries (and not all the twelve, as one would naturally expect). This is due to the nature of the numerator and denominator variables of the corresponding ${}_3F_2(1)$s.

To conclude, the van der Waerden ${}_3F_2(1)$ form of the 3-j coefficient is better than the other three forms due to Wigner, Racah, and Majumdar. This is because the van der Waerden form exhibits 12 of the 72 symmetries of the 3-j coefficient—these 12 being the Regge symmetries on which the classical symmetries are superposed. A set of six ${}_3F_2(1)$s is shown to be necessary and sufficient to account for all the symmetries of the 3-j coefficient.

References

Arima A 1999 Elliot's SU(3) model and its developments in nuclear physics *J. Phys. G: Nucl. Part. Phys.* **25** 581

Bailey W N 1935 *Generalized Hypergeometric Series*Cambridge Tracts in Mathematics and Mathematical Physics 32 (Cambridge: Cambridge University Press)

Biedernharn L C, Blatt J and Rose M E 1952 *Rev. Mod. Phys.* **24** 249

Biedenharn L C and Van Dam H 1965 *Quantum Theory of Angular Momentum* (New York: Academic)

Biedenharn L C and Louck J D 1981 *Angular Momentum in Quantum Physics: Theory and Application* Encyclopedia of Mathematics and its Applications vol 8

Biedenharn L C and Louck J D 1984 *The Racah Wigner Algebra in Quantum Theory*, Encyclopedia of Mathematics and its Applications vol 9

Bohr A and Mottelson B R 1969 *Nuclear Structure vol I: Single-Particle Motion; vol II: Nucelar Deformations* 1st edn (New York: Benjamin) Reprinted (1998, 1999, 2008) (Singapore: World Scientific)

Born M and Jordan P 1925 Zur Quantenmechanik *Z. Phys.* **34** 858–88 English translation, paper 13, 1968 *Sources of Quantum Mechanics* ed B I van der Waerden (New York: Dover).

Brink D M and Satchler G R 1962 *Angular Momentum* (Oxford: Clarendon)

Brown G E and Lee C-H 2006 *Hans Bethe and His Physics* (Singapore: World Scientific)

Condon E U and Shortley G H 1935 *The Theory of Atomic Spectra* (Cambridge: Cambridge University Press)

Crease R P and Mann C C 1996 *The second creation: Makers of the Revolution in Twentieth Century Physics* (New Brunswick, NJ: Rutgers University Press)

de-Shalit A and Talmi I 1963 *Theoretical Nuclear Physics: Nuclear Structure* (New York: Academic) (reprinted by Dover in 2004)

Edmonds A R 1957 *Angular Momentum in Quantum Mechanics* (Princeton, NJ: Princeton University Press)

Eisenberg J M and Greiner W 1975 *Nuclear Theory* vol I (New York: Holland)

Eisenberg J M and Greiner W 1976 *Nuclear theory*Volumes 1 and 2 (New York: North Holland)

Elliott J P and Lane A M 1957 *Angular Momentum in Quantum Mechanics* (Princeton, NJ: Princeton University Press)

Elliott J P 1958 *Proc. R. Soc. London* A **245** 562

Fano U and Racah G 1959 *Irreducible Tensorial Sets* (New York: Academic)

Feenberg E 1955 *Shell Theory of the Nucleus* (Princeton, NJ: Princeton University Press)

Gelfand L M, Minlos R A and Shapiro Z Y 1958 *Representation of the Rotation and Lorentz groups and Then Applications* (New York: Dover)

Gellmann M and Neeman Y 1964 *The Eightfold Way* (New York: Benjamin)

Hamermesh M 1962 *Group Theory and its Applications to Physical Problems* (New York: Dover)

Joffe D 2006 Obituary The Guardian, London, May 14

Johnson G 1999 *Strong Beauty - Murray Gell-Mann and the Revolution in 20th Century Physics* (New York: Alfred A. Knopf)

Johansson H T and Forssen C 2015 Fast and accurate evaluation of Wigner 3j, 6j, and 9j symbols using prime factorization and multi-word integer arithmetic, ArXiv: 104.08329vs [comp. comp-ph]

Majumdar S D 1958 *Prog. Theor. Phys.* **20** 798

Mayer von M, Goeppert and Jensen J H D 1955 *Elementary Theory of Nuclear Shell Structure* (New York: Wiley)

Mehra J and Milton A 2000 *SClimbing the Mountain: The Scientific Biography of Julian Schwinger* (Oxford: Oxford University Press)

Moore W 2017 *Schrodinger: Life and thought* (Cambridge: Cambridge University Press)

Racah G 1942a *Phys. Rev.* **61** 186

Racah G 1942b *Phys. Rev.* **62** 438

Racah G 1943 *Phys. Rev.* **63** 367

Racah G 1949 *Phys. Rev.* **76** 1352

Raynal J 1978 *J. Math. Phys.* **19** 467–76

Regge T 1958 *Nuovo Cimento* **10** 544

Rose M E 1957 *Elementary Theory of Angular Momentum* (New York: Wiley)

Rose M E 1955 *Mulitpole Fields* (New York: Wiley)

Rotenberg M, Bivins R, Mctropolis N and Wooten J K 1959 *The 3-j and 6-j Symbols* (Cambridge, MA: MIT Press)

Schrödinger E R J A 1887–1961 *Nobel Lectures, Physics 1922–1941*, 'Erwin Schrödinger Biography' from Nobel Prize.org

Sen D 2014 *Curr. Sci.* **107** 203

Smorodinskii Ya A and Shelepin Leonid A 1972 *Sov. Phys. Uspekhi* **15** 1–24

Srinivasa Rao K 1978 *J. Phys.* **11A** L69

Srinivasa Rao K 1990 New results in quantum theory of angular momentum *Fourth Regional Conf. in Mathematical Physics (Tehran)*

Srinivasa Rao K and Rajeswari V 1993 *Selected Topics in Quantum Theory of Angular Momeutm* (New York: Springer)

Stanton A and Wigner E P 1995 *The recollections of Eugene P Wigner* (NewYork: Springer)

Varshalovich D A, Moskalev A N and Khersonskii V K 1975 *A Quantum Theory of Angular Momentum*[in Russian] (Leningrad: Nauka)

Weyl H 2014 *The Theory of Groups and Quantom Mechanics*Trans. ed H P Robertson (New York: Martino Five Books) Reprint of 1931 edition

Whipple F J W 1925 *Proc. London Math. Soc.* **23** 104–14

Wigner E P 1927 Über die Erhaltungssätze in der Quantenmechanik, Nachrichten von der Gesellschaft der Wissenschaften.

Wigner E P 1959 *Group Theory and its Application to the Quantum Mechanics of Atomic Spectra*Translation from German by (1902–95) ed J J Griffin. (New York: Academic)

Yutsis A P, Levinson I B and Vanagas V V 1960 *The Theory of Angular Momentum* (Vilnius: The Israel Program for Scientific Translations)

Yutsis A P and Bandzaitis A A 1977 *The Theory of Angular Momentum in Quantum Mechanics* [in Russian] (Vilnius: Mokslas)

Zare R N 1988 *Angular Momentum* (New York: Wiley)

Zweig G 1980 *Orgins of the Quark Model*CALT-68-805 (Pasadena, CA: California Institute of Technology)

Chapter 6

Angular momentum recoupling and sets of ${}_4F_3(1)$s

6.1 Introduction: historical

In this chapter, we introduce the concept of the recoupling of angular momenta. For for the case of three angular momenta, the recoupling coefficient was introduced by Giulio Racah, renowned as the Racah coefficient, or the 6-j coefficient. As noted in the previous chapter, Racah, in a celebrated series of four papers (1942–1949) on the *Theory of Complex Spectra*, systematically developed algebraic techniques and provided powerful tools for the recoupling scheme of evaluating the scalar product of tensor operators. He introduced the coefficient as a transformation coefficient between alternate coupling for the addition of three angular momenta. We recognize the fundamental pioneering contributions made by Wigner and Racah to the coupling and recoupling of two and three angular momenta, respectively, as the basic elements of quantum theory of angular momentum which is referred to in the literature as *The Racah–Wigner Algebra* (Biedenharn and Louck 1984). The *classical* aspects of quantum theory of angular momentum were considered complete until Tulio Regge in two short communications (1958, 1959) showed that the symmetry groups of the 3-j and the 6-j coefficients are larger groups containing 72 and 144 elements, respectively, rather than the *classical* 12 and 24 element groups. These discoveries were the starting points of new developments in quantum theory of angular momentum, such as the generalization of Clebsch–Gordan coefficients to arbitrary complex arguments—an essential extension when in connection with Regge poles and trajectories (1959, 1960).

In this context, we wish to draw the attention of the interested reader to a period piece—an excellent review article of Smorodinskii and Shelepin (1972)—in which the close relation of the theory of Clebsch–Gordan coefficients is drawn with combinatorics, finite differences, special functions, complex angular momenta, projective and multi-dimensional geometry, and several other branches of mathematics. It is an inspiring article that will benefit interested researchers.

doi:10.1088/978-0-7503-1496-1ch6

In section 6.2, the recoupling coefficient of the Racah, or 6-j, coefficient, which occurs in the coupling of three angular momenta, is defined. Section 6.3 is on the Regge symmetries of the Racah coefficient. In section 6.4, we present two equivalent sets of ${}_4F_3(1)$s, which are necessary and sufficient to account for the known 144 symmetries of the 6-j coefficient. In section 6.5, the inter-relationship between these two sets of ${}_4F_3(1)$s is established. Section 6.6 is about the Bargmann–Shelepin arrays for the 6-j coefficient. In section 6.7, the Bailey transformation, when recursively used, is shown to result in a set of 20 transformations. The symmetries of the 6-j coefficient are discussed in terms of the basis states in section 6.8. In section 6.9, we present the generators for the 144 symmetries of the 6-j coefficient, and the last section presents a summary and additional remarks.

6.2 Addition of three angular momenta

Consider the coupling of three angular momenta:

$$\vec{j_1} + \vec{j_2} + \vec{j_3} = \vec{J}. \tag{6.1}$$

In terms of the coupling of two angular momenta detailed in chapter 5, we can couple any two of the three angular momenta to form an intermediate angular momentum, which can then be coupled to the third angular momentum to form the final angular momentum $\vec{J}$. Two of the three possible additional schemes are thus:

$$\begin{aligned} \vec{j_1} + \vec{j_2} &= \vec{j_{12}}, \qquad \vec{j_{12}} + \vec{j_3} = \vec{J}, \\ \vec{j_2} + \vec{j_3} &= \vec{j_{23}}, \qquad \vec{j_1} + \vec{j_{23}} = \vec{J}, \end{aligned} \tag{6.2}$$

where $\vec{j_{12}}$ and $\vec{j_{23}}$ are the intermediate coupled angular momenta. The complete set of commuting generators for these two coupling schemes are then

$$j_!^2, j_2^2, j_{12}^2, j_3^2, J^2, J_z \qquad \text{and} \qquad j_!^2, j_2^2, j_3^2, j_{23}^2, J^2, J_z. \tag{6.3}$$

These then provide the labels for the orthonormal basis vectors which are denoted as

$$|(j_1 j_2)j_{12}\ j_3\ J\ M\rangle \qquad \text{and} \qquad |j_1\ (j_2 j_3)j_{23}\ J\ M\rangle, \tag{6.4}$$

respectively. It is straightforward to show using (5.119), that the coupled state $|(j_1 j_2)\, j_{12}\, j_3\ J\ M\rangle$ can be decoupled into a sum over the product of two Clebsch–Gordan coefficients, multiplied by the product of three uncoupled states $|j_1\ m_1\rangle|j_2\ m_2\rangle\, j_3\ m_3\rangle$, and then using the inverse of (5.119), viz (5.121), we can recouple these three states to get the other recoupled state: $|j_1\ (j_2 j_3)j_{23}\ J\ M\rangle$. The orthonormal basis states (6.4) are thus related to the orthogonal transformation

$$\begin{aligned} |(j_1 j_2)\ j_{12}\ j_3\ J\ M\rangle = \sum_{j_{23}} & U(j_1\, j_2\ J\ j_3;\ j_{12}\, j_{23}) \\ & |j_1\ (j_2 j_3)\ j_{23}\ J\ M\rangle \end{aligned} \tag{6.5}$$

or, equivalently:

$$|j_1\,(j_2 j_3)\;j_{23}\;J\;M\rangle = \sum_{j_{12}} U(j_1\;j_2\;J\;j_3;\;j_{12}\,j_{23})\;|(j_1 j_2)\;j_{12}\;j_3\;J\;M\rangle \tag{6.6}$$

where $U(j_1\,j_2\,J\,j_3;j_{12}\,j_{23})$ is the recoupling coefficient, which, when written explicitly, is the product of four Clebsch–Gordan coefficients summed over two (of the six) independent projection quantum numbers

$$U(j_1\;j_2\;J\;j_3;\;j_{12}\,j_{23}) = \sum_{m_1,m_2} C(j_1\;j_2\;j_{12};\,m_1\,m_2\,m_{12}) \times C(j_{12}\;j_3\;J;\,m_{12}\,m_3\,M)\;C(j_2\;j_3\;j_{23};\,m_2\,m_3\,m_{23})\;C(j_1\;j_{23}\;J;\,m_1\,m_{23}\,M). \tag{6.7}$$

The Racah coefficient is related to the recoupling coefficient as

$$W(j_1\;j_2\;J\;j_3;\;j_{12}\,j_{23}) = [(2\,j_{12}+1)\,(2\,j_{23}+1)]^{-\frac{1}{2}}\;U(j_1\;j_2\;J\;j_3;\;j_{12}\,j_{23}) \tag{6.8}$$

and it exhibits the symmetry properties more elegantly. Racah's achievement (1942a) was to show that this recoupling coefficient can be written as a single sum, independent of m_i; the projection quantum numbers are

$$W(abcd;\,ef) = \sum_P (-1)^P (P+1)!\left\{\prod_{i=1}^{4}(P-\alpha_i)!\prod_{j=1}^{3}(\beta_j - P)!\right\} \tag{6.9}$$

with

$$N = (-1)^{a+b+c+d}\Delta(abe)\;\Delta(cde)\;\Delta(acf)\;\Delta(bdf) \tag{6.10}$$

and

$$\begin{gathered}\alpha_1 = a+b+e,\ \alpha_2 = c+d+e,\\ \alpha_3 = a+c+f,\ \alpha_4 = b+d+f,\\ \beta_1 = a+b+c+d,\ \beta_2 = a+d+e+f,\\ \beta_3 = b+c+e+f,\end{gathered} \tag{6.11}$$

$$P_{\min} \leqslant P \leqslant P_{\max}, \tag{6.12}$$

$$P_{\min} = \max(\alpha_1, \alpha_2, \alpha_3, \alpha_4), \qquad P_{\max} = \min(\beta_1, \beta_2, \beta_3) \tag{6.13}$$

$$\Delta(xyz) = \left[\frac{(-x+y+z)!\,(x-y+z)!\,(x+y-z)!}{(x+y+z+1)!}\right]^{\frac{1}{2}}, \tag{6.14}$$

and the property of the factorials in the numerator of (6.14), whose arguments must be non-negative, reiterates the property of the triangle inequality

$$|x-y| \leqslant z \leqslant (x+y). \tag{6.15}$$

Note that, for ease of typesetting, we have chosen to use the simpler lower case roman letters *a*, *b*, *c*, *d*, *e*, *f* instead of the angular momenta j_1, j_2, J, j_3, j_{12}, j_{23}.

6.3 Symmetries of the Racah coefficient

The symmetries of the Racah coefficient are more easily interpreted in terms of the 6-*j* coefficient (or 6-*j* symbol) by introducing a phase factor in addition to the numerical factors that relate the $U(a, b, c, d, e, f)$-coefficient to the *W*-coefficient (6.8) as

$$\begin{Bmatrix} a & b & e \\ d & c & f \end{Bmatrix}, \tag{6.16}$$

and they are stated as invariance of the series (6.9) to:

(i) the three column permutations of the 6-*j* symbol, and
(ii) invariance to the interchange of any two elements in the first two with the corresponding elements of the second row (which we will refer to as 'row' permutations)

Explicitly, (i) states that

$$\begin{aligned} \begin{Bmatrix} a & b & e \\ d & c & f \end{Bmatrix} &= \begin{Bmatrix} b & e & a \\ c & f & d \end{Bmatrix} = \begin{Bmatrix} e & a & b \\ f & d & c \end{Bmatrix} = \\ &= \begin{Bmatrix} b & a & e \\ c & d & f \end{Bmatrix} = \begin{Bmatrix} e & b & a \\ f & c & d \end{Bmatrix} = \begin{Bmatrix} a & e & b \\ d & f & c \end{Bmatrix}. \end{aligned} \tag{6.17}$$

Similarly, (ii) states that

$$\begin{aligned} \begin{Bmatrix} a & b & e \\ d & c & f \end{Bmatrix} &= \begin{Bmatrix} d & c & e \\ a & b & f \end{Bmatrix} \\ &= \begin{Bmatrix} d & b & f \\ a & c & e \end{Bmatrix} = \begin{Bmatrix} a & c & f \\ d & b & e \end{Bmatrix}. \end{aligned} \tag{6.18}$$

These are the *classical tetrahedral* symmetries of the 6-*j* coefficient, which are manifest when we note that the series is invariant to the four permutations of α_1, α_2, α_3, α_4 variables and to the three permutations of the β_1, β_2, β_3 variables—that is, the symmetries correspond to the symmetries of the S_4 and S_3 groups. The total number of symmetries of the 6-*j* coefficient are thus $24 \times 6 = 144$, a realization that should have dawned on Regge, who found six additional symmetries to the 24 symmetries;

the tetragonal symmetries were the only ones that were known till then. Regge also wrote these down explicitly as

$$\begin{Bmatrix} a & b & e \\ d & c & f \end{Bmatrix} = \begin{Bmatrix} a & \frac{1}{2}(b+c+e-f) & \frac{1}{2}(b-c+e+f) \\ d & \frac{1}{2}(b+c-e+f) & \frac{1}{2}(-b+c+e+f) \end{Bmatrix} \tag{6.19}$$

$$= \begin{Bmatrix} \frac{1}{2}(a-d+e+f) & b & \frac{1}{2}(a+d+e-f) \\ \frac{1}{2}(-a+d+e+f) & c & \frac{1}{2}(a+d-e+f) \end{Bmatrix} \tag{6.20}$$

$$= \begin{Bmatrix} \frac{1}{2}(a+b+c-d) & \frac{1}{2}(a+b-c+d) & e \\ \frac{1}{2}(-a+b+c+d) & \frac{1}{2}(a-b+c+d) & f \end{Bmatrix} \tag{6.21}$$

$$= \begin{Bmatrix} \frac{1}{2}(b+c+e-f) & \frac{1}{2}(a-d+e+f) & \frac{1}{2}(a+b-c+d) \\ \frac{1}{2}(b+c-e+f) & \frac{1}{2}(-a+d+e+f) & \frac{1}{2}(a-b+c+d) \end{Bmatrix} \tag{6.22}$$

$$= \begin{Bmatrix} \frac{1}{2}(b-c+e+f) & \frac{1}{2}(a+d+e-f) & \frac{1}{2}(a+b+c-d) \\ \frac{1}{2}(-b+c+e+f) & \frac{1}{2}(a+d-e+f) & \frac{1}{2}(-a+b+c+d) \end{Bmatrix}. \tag{6.23}$$

It is important to note that the form (6.9) can be found for the first time only in the short communication of Regge (1959). Racah (1942a, 1942b) dealt with only the series obtained by substituting in (6.9)

$$s = \beta_1 - P = a + b + c + d - P \tag{6.24}$$

resulting in a series representation for the Racah coefficient, which, when rearranged into the generalized hypergeometric function of unit argument, is a ${}_4F_3(ABCD;\, EFG;\, 1)$ with

$$\begin{gathered} A = e - a - b, \quad B = e - c - d, \quad C = a - c - f, \\ D = f - b - d, \quad E = -a - b - c - d - 1, \quad F = e + f - a - d + 1, \\ G = e + f - b - c + 1. \end{gathered} \tag{6.25}$$

This is the series given in the appendix of Rose's (1955) monograph, titled *Multipole Fields.* The reader may think that the inherent property of the trivial $S_4 \times S_3$ symmetries, due to the permutation of the four αs and the three βs, would yield the $24 \times 6 = 144$ symmetries of the of the 6-j coefficient—viz the six Regge symmetries on which are superposed the 24 *classical tetrahedral symmetries.* One of us (Srinivasa Rao 1978) found that the nature of the numerator and denominator variables plays a vital role in relating the permutations of the numerator–denominator variables of the ${}_4F_3(1)$ to the symmetries of the Racah or 6-j coefficient. It follows from the triangle inequalities satisfied by the triads in

$$N = \Delta(abe)\ \Delta(cde)\ \Delta(acf)\ \Delta(bdf) \tag{6.26}$$

that

$$\begin{aligned} e - a - b \leqslant 0, \quad & e - c - d \leqslant 0, \\ a - c - f \leqslant 0, \quad & b - d - f \leqslant 0, \end{aligned} \tag{6.27}$$

are all negative integer quantities. However, the triangle inequalities do not say anything about the nature of the relative sums of the elements in the columns of the 6-j coefficient. We have to therefore study this aspect in detail.

6.4 Two sets of ${}_4F_3(1)$s

By setting—as in (6.9)—$s = \beta_k - \mathrm{P}$, $k = 1, 2, 3$, in succession, we get

$$\begin{Bmatrix} a & b & e \\ d & c & f \end{Bmatrix} = N'(-1)^{\beta_k} \sum_k (-1)^k (\beta_k - s + 1)! \times \left[\prod_{i=1}^{4} (\beta_k - \alpha_i - s)! \prod_{j=1}^{3} (s - \beta_k + \beta_j) \right]^{-1}, \tag{6.28}$$

where $N' = N(-1)^{a+b+c+d}$ and N is given by (6.26). We call this set I of three ${}_4F_3(1)$ series representations for the 6-j coefficient. Only one member of this set I, corresponding to $s = \beta_1 - P$, can be found in the literature (i.e. Biedenharn and Louck 1981, Varshalovich *et al* 1975). In 1975, Srinivasa Rao, Santhanam, and Venkatesh pointed out that the set I of three ${}_4F_3(1)$s is *necessary and sufficient* to account for the 144 symmetries of the 6-j coefficient. For a given series belonging to the set I of three series, representations can account for only 48 of the 144 symmetries due to the permutation of all four αs, but only two of the βs, since, in (6.28), one of the three βs appears in the numerator.

In (6.28), setting $s = P - \alpha_\ell$, $\ell = 1, 2, 3, 4$, in succession, Srinivasa Rao and Venkatesh (1978) obtained the following set II of four series representations for the 6-j coefficient

$$\begin{Bmatrix} a & b & e \\ d & c & f \end{Bmatrix} = N'(-1)^{\alpha_\ell} \sum_s (-1)^s (\alpha_\ell + s + 1)! \times \left[\prod_{i=1}^{4} (s + \alpha_\ell - \alpha_i)! \prod_{j=1}^{3} (\beta_j - \alpha_\ell - s)! \right]^{-1}, \tag{6.29}$$

where a series belonging to set II of ${}_4F_3(1)$s exhibits only 36 of the 144 symmetries arising, due to the permutation of all the three βs, but only due to three of the four αs, since α_1 is now in the numerator of (6.29).

As in the case of rearranging the 3-j coefficient as a set of six ${}_3F_2(1)$s, we use the relations

$$n! = \Gamma(n-1) \quad \text{and} \quad \Gamma(1-z-n) = (-1)^n \frac{\Gamma(z)\Gamma(1-z)}{\Gamma(z+n)} \tag{6.30}$$

to get from (6.28) one member of the set I of three ${}_4F_3(1)$s:

$$\begin{aligned} \begin{Bmatrix} a & b & e \\ d & c & f \end{Bmatrix} = (-1)^{E+1} N\, \Gamma(1-E) \\ \times [\Gamma(1-A, 1-B, 1-C, 1-D, F, G)]^{-1} \\ \times {}_4F_3(ABCD;\, EFG;\, 1)(-1)^k \end{aligned} \tag{6.31}$$

where

$$\begin{aligned} A = e-a-b, \quad B = e-c-d, \quad C = f-a-c, \\ D = f-b-d, \quad E = -a-b-c-d-1, \\ F = e+f-b-c+1, \\ G = e+f-a-d+1. \end{aligned} \tag{6.32}$$

For $k = 2$ and 3, we get for the numerator and denominator variables,

$$\begin{aligned} A = a-b-e, \quad B = d-c-e, \quad C = a-c-f, \\ D = d-b-f, \quad E = -b-c-e-f-1, \\ F = a+d-b-c+1, \\ G = a+d-e-f+1. \end{aligned} \tag{6.33}$$

$$\begin{aligned} A = b-a-e, \quad B = c-d-e, \quad C = c-a-f, \\ D = b-d-f, \quad E = -a-d-e-f-1, \\ F = b+c-a-d+1, \\ G = b+c-e-f+1. \end{aligned} \tag{6.34}$$

Obviously, superposing the column permutations of the 6-j coefficient on the variables of the ${}_4F_3(1)$ in (6.28) yields the set I of three ${}_4F_3(1)$s: (6.30)–(6.33). We note that the superposition of the 'row' permutations ('row' permutation defined here as the interchange of the two elements of the first row with the corresponding two elements of the second row of the 6-j coefficient) of the ${}_4F_3(1)$ in (6.30) results only in a permutation of the numerator and denominator variables amongst themselves in the ${}_4F_3(1)$ set.

The reason why this single ${}_4F_3(1)$ exhibits only 48 of the 144 symmetries is due to the nature of the numerator and denominator variables belonging to this set I of ${}_4F_3(1)$s, which satisfy the Saalschütz condition (Slater 1966):

$$A + B + C + D + 1 = E + F + G. \tag{6.35}$$

Here, we notice that, for all the physical values of a, b, c, d, e, f, the numerator variables A, B, C, $D \leqslant 0$ are negative integers and the denominator variable $E \leqslant -1$ is negative and satisfies the condition

$$(|A|, |B|, |C|, \text{ or } |D|) < |E| \tag{6.36}$$

by virtue of the four triangular inequalities to be satisfied by the triads in (6.26). However, since the triangle inequalities do not give any information about the relative magnitudes of the column sums of the 6-j coefficient, F and G can be either positive or zero. For the ${}_4F_3(1)$ to be convergent, there must be a numerator variable such that

$$(|A|, |B|, |C|, |D|) < (|F|, |G|). \tag{6.37}$$

However, a comparison of the denominator variables with the numerator variables, along with the triangular conditions, requires the inequality to be satisfied by them as

$$(|F| \text{ or } |G|) < (|A|, |B|, |C| \text{ or } |D|) \tag{6.38}$$

in all of the three ${}_4F_3(1)$ belonging to set I. From (6.37) and (6.38), it follows that $F > 0$ and $G > 0$ for the hypergeometric series to be convergent. Or, in other words, the ${}_4F_3(1)$ belonging to set I are well defined and convergent, if and only if

$$e + f \geqslant a + d \quad \text{and} \quad e + f \geqslant b + c. \tag{6.39}$$

Obviously, when either

$$a = b = c = d = e = f \quad \text{or} \quad a + d = e + f = b + c, \tag{6.40}$$

all three ${}_4F_3(1)$s are convergent and are equal to either

$$\begin{gathered} {}_4F_3(-a, -a, -a, -a; -4a - 1, 1, 1; 1) \\ \text{or} \\ {}_4F_3(e - a - b, a - f - c, f - a - c, a - b - e; -2b - c - e, 1, 1; 1). \end{gathered} \tag{6.41}$$

For the other physically allowed values of a, b, c, d, e, f, only one or two of the set I of three ${}_4F_3(1)$s is convergent. Thus, the set I of three ${}_4F_3(1)$s is necessary and sufficient to account for the 144 symmetries of the 6-j coefficient.

If the domain of definition of the 6-j coefficient is specified by the four triads in (6.10), then the sub-domains specified by (6.39), and the corresponding sub-domains for the remaining two ${}_4F_3(1)$s are specified by

$$\begin{aligned} a+d &\geqslant e+f, \quad a+d \geqslant b+c \quad \text{and} \\ b+c &\geqslant e+f, \quad b+c \geqslant a+d. \end{aligned} \tag{6.42}$$

Thus, we have overlapping regions such that in:
region (I), any one of the three members of the set I are defined;
region (II), any two of the three members of the set I are defined; and
region (III) all the three members of the set I are defined.

Minton (1970) announced a 'new' symmetry for the Racah coefficient. Soon after, Yakimiw (1971) pointed out that the 'new' symmetry violates the triangle inequalities. However, the question of why it was a valid procedure for Minton to use a transformation formula for the ${}_4F_3(1)$, which resulted in an invalid result, was unanswered. This was precisely the point answered by Srinivasa Rao, Santhanam, and Venkatesh (1975), which was based on the following observations.

The Bailey transformation between two terminating Saalschützian ${}_4F_3(1)$ series' (Slater 1966, p 64, equation (2.4.1.7)) is given by:

$$\begin{aligned} &{}_4F_3(A, B, C, D; E, F, G; 1) \\ &= \Gamma(E+F-A-B-D, F-C-D, F, E+F-A-B-C) \\ &\times [\Gamma(E+F-A-B, E=F-A-B-C, F-C, F-D)]^{-1} \\ &\times {}_4F_3(E-A, E-B, C, D; E, E+F-A-B, \\ &E+G-A-B; 1) \end{aligned} \tag{6.43}$$

where we have used the notation for the product of Gamma functions (5.140). The above relation is invariant to the interchange of A and B, and C and D. Srinivasa Rao, Santhanam, and Venkatesh (1975) observed that if the variable is $C = 0$, then the Γ functions in (6.43) get cancelled to become just one, and the ${}_4F_3(1)$ series on the right and left sides of (6.43) become equal to one. Choosing the variables of ${}_4F_3(1)$ given by (6.32), the identity obtained was

$$\begin{aligned} &{}_4F_3(e-a-b, e-c-d, 0, a-b+c-d; \\ &\quad -a-b-c-d-1, c-d+e+1, a-b+e+1; 1) = 1 \\ &\quad = {}_4F_3(-e-1-a-b, -e-1-c-d, 0, a-b+c-d; \\ &\quad -a-b-c-d-a, c-d-e, a-b-e). \end{aligned} \tag{6.44}$$

An identification of the corresponding numerator and denominator variables on the left- and right-hand sides of the above identity (6.44) clearly shows that the Bailey transformation—when applied to the ${}_4F_3(1)$ series with the variables given by (6.32) for the Racah coefficient—leads to the simple substitution: $e \to -e-1$. This,

$$j \to -j-1, \tag{6.45}$$

is an inherent mathematical symmetry, since it leaves the basic angular momentum relations, (5.70) and (5.72), invariant. Explicitly, the factors

$$j(j+1) \quad \text{and} \quad [(j \mp m)(j \pm m + 1)]^{\frac{1}{2}} \tag{6.46}$$

are invariant to (6.45).

Though physically, negative angular momenta are not relevant or significant, this property (6.45) is called Yutsis 'mirror' symmetry. Unphysical transformations of this type have been extensively studied by Yutsis and Bandzaitis (1977). An extension of the Wigner 3-j and the Racah 6-j coefficients to values of the representation parameters of $SU(2)$, related to the usual ones by (6.45), have been studied by Bandzaitis *et al* (1964). Through analytic continuation in these parameters, a deep connection between the corresponding Wigner coefficients has been pointed out by Holman and Biedenharn (1966, 1968).

Beyer, Louck, and Stein (1987) used the Bailey transformation (6.43) for the terminating Saalschützian ${}_4F_3(1)$ series (Bailey 1935, p 56) to study the symmetry group S_5 of two-term relations for the ${}_4F_3(1)$ series. Using the relation between the 6-j coefficient and the terminating Saalschützian ${}_4F_3(1)$, and applying it to the new symmetry group, Louck *et al* (1986) showed that the group of 144 symmetries are extended to a group of 23 040 symmetries by extending the domain of these coefficients. Clearly, this extended domain contains also 'unphysical' arguments for the 6-j coefficient. This extended symmetry group of the order 23 040 has also been encountered by D'Adda *et al* (1971, 1974) in their unified treatment of $SU(2)$ and $SU(1,1)$ 6-j coefficients.

Rearranging (6.29), Srinivasa Rao, and Venkatesh (1978) obtained one member of the following set II of four ${}_4F_3(1)$s for the 6-j coefficient:

$$\begin{Bmatrix} a & b & e \\ d & c & f \end{Bmatrix} = (-1)^{A'-2} N\, \Gamma(A') \times [\Gamma(1-B', 1-C', 1-D', E', F', G')]^{-1} \times {}_4F_3(A'B'C'D'; E'F'G'; 1) \tag{6.47}$$

where

$$\begin{aligned} A' &= a+b+e+2, \quad B' = a-c-f, \quad C' = b-d-f, \\ D' &= e-c-d, \quad E' = a+b-c-d+1, \\ F' &= e+f-b-c+1, \\ G' &= e+f-a-d+1. \end{aligned} \tag{6.48}$$

For $\ell = 2, 3, 4$, we get, for the numerator and denominator variables, the sets:

$$\begin{aligned} A' &= c+d+e+2, \quad B' = c-a-f, \quad C' = d-b-f, \\ D' &= e-a-b, \quad E' = c+d-a-b+1, \\ F' &= c+e-b-f+1, \\ G' &= d+e-a-f+1. \end{aligned} \tag{6.49}$$

$$\begin{gathered} A' = a + c + f + 2, \quad B' = c - d - e, \quad C' = a - b - e, \\ D' = f - b - d, \quad E' = a + c - b - d + 1, \\ F' = a + f - d - e + 1, \\ G' = c + f - b - e + 1. \end{gathered} \tag{6.50}$$

$$\begin{gathered} A' = b + d + f + 2, \quad B' = b - a - e, \quad C' = d - c - e, \\ D' = f - a - c, \quad E' = b + d - a - c + 1, \\ F' = b + f - c - e + 1, \\ G' = d + f - a - e + 1. \end{gathered} \tag{6.51}$$

Obviously, the set of parameters of the four ${}_4F_3(1)$s is spanned by superposing on the 'row' permutations of the 6-j symbol on the parameters (6.34). Superposing the column permutations of a 6-j coefficient on a given ${}_4F_3(1)$ belonging to this set II results only in a permutation of the numerator and denominator variables among themselves.

For the set II of ${}_4F_3(1)$s, the nature of the numerator variables is: while A' is a positive integer, the other three variables B', C', D' are negative integers. All three denominator variables E', F', G' must be positive if the ${}_4F_3(1)$ is to be convergent and well-defined.

As in the case of the set I of ${}_4F_3(1)$s, we can show that for the set II of ${}_4F_3(1)$s, the domains of definition of the four sets of variables, given in (6.48)–(6.51), are

$$a + b \geqslant c + d, \quad a + e \geqslant d + f, \quad b + e \geqslant c + f, \tag{6.52}$$

$$c + d \geqslant a + b, \quad c + e \geqslant b + f, \quad d + e \geqslant a + f, \tag{6.53}$$

$$a + c \geqslant b + d, \quad a + f \geqslant d + e, \quad c + f \geqslant b + e, \tag{6.54}$$

$$b + d \geqslant a + c, \quad b + f \geqslant c + e, \quad d + f \geqslant a + e. \tag{6.55}$$

These are four overlapping sub-domains of the domain of definition for the 6-j coefficient. Shown in the figure below are the domains of definitions for the sets I and II of ${}_4F_3(1)$s. There are regions in which only one, two, or three of the set of four ${}_4F_3(1)$s are defined. Only when

$$a = b = c = d = e = f \qquad \text{or} \qquad a = d, b = e, c = f, \tag{6.56}$$

are all four ${}_4F_3(1)$s well defined. Thus, the set II of four ${}_4F_3(1)$s is necessary and sufficient to account for all the known 144 symmetries of the 6-j coefficient—each member of this set accounts for only 36 of the 144 symmetries. These are due to the permutations of the three variables of the numerator and the three variables of the denominator—$S_3 \times S_3$.

6.5 Inter-relationship of the two sets of ${}_4F_3(1)$s

The answers to the questions of whether these two equivalent sets of ${}_4F_3(1)$s are necessary and sufficient to represent the 6-j coefficient as a generalized hypergeometric function of unit argument are related to one another. And the answer in the affirmative for this question is due to Srinivasa Rao and Rajeswari (1985). Since more than one variable in sets I and II is a negative integer, we can generalize the property of reversal of the series given by Bailey (1935) for the case of a ${}_3F_2(1)$ to obtain the identity

$$\begin{aligned}{}_4F_3(1bcd;\, EFG;\, 1) &= (-1)^D \\ &\times\Gamma(1-A,\, 1-C,\, 1-D,\, F,\, G,\, D-E+1) \\ &\times[\Gamma(D-A+1,\, D-B+1,\, D-C+1,\, 1-E,\, F-D,\, G-D)]^{-1} \\ &\times {}_4F_3(A',\, B',\, C',\, D';\, E'F'G';\, 1)\end{aligned} \tag{6.57}$$

where $D(=D')$ is the minimum of the negative numerator variables that determine the number of terms in the series. If we denote by ξ the column vector of the variables of the ${}_4F_3(1)$ on the left-hand side of (6.57), viz $(A, B, C, D+1, E, F, G)$, and by ξ' the column vectors of the variables on the right-hand side, viz $(A', B'C', D'+1, E', F', G')$, then

$$\xi' = t\,\xi \tag{6.58}$$

where the transforming matrix is

$$t = \begin{bmatrix} 0 & 0 & 0 & 1 & -1 & 0 & 0 \\ 0 & 0 & 0 & 1 & 0 & -1 & 0 \\ 0 & 0 & 0 & 1 & 0 & 0 & -1 \\ 0 & 0 & 0 & 1 & 0 & 0 & 0 \\ -1 & 0 & 0 & 1 & 0 & 0 & 0 \\ 0 & -1 & 0 & 1 & 1 & 0 & 0 \\ 0 & 0 & -1 & 1 & 0 & 1 & 0 \end{bmatrix}. \tag{6.59}$$

Using (6.57) in (6.31) after some simple algebraic manipulations, we get

$$\begin{aligned}\begin{Bmatrix} a & b & e \\ d & c & f \end{Bmatrix} &= (-1)^{A'-2}N\,\Gamma(A') \\ &\times[\Gamma(1-B',\, 1-C',\, 1-D',\, E',\, F',\, G')]^{-1} \\ &\times {}_4F_3(A'B'C'D';\, E'F'G';\, 1),\end{aligned} \tag{6.60}$$

which is a ${}_4F_3(1)$ belonging to set II, except for a permutation amongst the numerator and denominator variables, and the same with (6.47).

While using (6.57) in (6.31), one has to choose one of the numerator variables to be D, which determines the number of terms D can be any one of the four numerator variables of the ${}_4F_3(1)$s belonging to set I. Thus, from any one of the three ${}_4F_3(1)$s of set I, by the *reversal of series*, all four ${}_4F_3(1)$s of set II can be obtained. By a similar argument, any one of the four ${}_4F_3(1)$s of the set I can be obtained.

6.6 Bargmann–Shelepin arrays

A notation due to Bargmann (1962) and Shelepin (1964) expresses the 6-j coefficient as a standard 4×3 symbol

$$\begin{Bmatrix} a & b & e \\ d & c & f \end{Bmatrix} = \left\| \begin{matrix} \beta_1 - \alpha_1 & \beta_2 - \alpha_1 & \beta_3 - \alpha_1 \\ \beta_1 - \alpha_2 & \beta_2 - \alpha_2 & \beta_3 - \alpha_2 \\ \beta_1 - \alpha_3 & \beta_2 - \alpha_3 & \beta_3 - \alpha_3 \\ \beta_1 - \alpha_4 & \beta_2 - \alpha_4 & \beta_3 - \alpha_4 \end{matrix} \right\| = \|R_{ik}\|, \tag{6.61}$$

which is invariant to four row permutations and three column permutations. The elements of $\|R_{ik}\|$ satsify the 18 relations

$$\begin{aligned} R_{kk} + R_{mn} &= R_{kn} + R_{mk} \\ R_{4k} + R_{mn} &= R_{4n} + R_{mk} \end{aligned} \tag{6.62}$$

for all $k \neq m$, $k \neq n$ and $k, m, n = 1, 2, 3$. Equivalently, we can state that all the 2×2 minors of $\|R_{ik}\|$, viz

$$\begin{vmatrix} R_{ij} & R_{ik} \\ R_{\ell j} & R_{\ell k} \end{vmatrix} \tag{6.63}$$

obey $R_{ij} + R_{\ell k} = R_{ik} + R_{\ell j}$, where $ij, k, \ell = (1, 2, 3, 4)$, such that there is no 2×2 minor with the column (or second index) being four, since the Bargmann–Shelepin array is a rectangular 4×3 array.

It has been shown (Srinivasa Rao and Rajeswari 1989) that the numerator and denominator variables of the set I of ${}_4F_3(1)$s given by (6.31) can also be written in terms of the elements of the standard Bargmann–Shelepin symbol $\|R_{ik}\|$ as

$$\begin{aligned} &A = -R_{1p}, \quad B = -R_{2p}, \quad C = -R_{3p}, \quad D = -R_{4p} \\ &E = -R_{1p} - R_{2p} - R_{3q} - R_{4r} - 1, \quad F = R_{3q} - R_{3p} + 1, \\ &G = R_{4r} - R_{4p} + 1, \end{aligned} \tag{6.64}$$

for $(pqr) = (123)$ cyclically, using (6.62). It is now possible to express the 4×3 symbol in terms of the numerator and denominator variables of the set I of ${}_4F_3(1)$s, using (6.62) as

$$\begin{Bmatrix} a & b & e \\ d & c & f \end{Bmatrix} = \left\| \begin{matrix} -A & F - A - 1 & G - A - 1 \\ -B & F - B - 1 & G - B - 1 \\ -C & F - C - 1 & G - C - 1 \\ -D & F - D - 1 & G - D - 1 \end{matrix} \right\| = \|R_{ik}\| \tag{6.65}$$

where the negative variable E does not appear in (6.65) and the non-standard 4×3 symbol exhibits only 48 of the 144 symmetries due to its manifest invariance to four row permutations and two column permutations. Since there exists a set I of numerator and denominator variables of the ${}_4F_3(1)$s due to the cyclic permutations of $(pqr) = (123)$ in (6.64), it follows that there exists a set I of three non-standard 4×3 symbols equivalent to the standard Bargmann–Shelepin symbol, which is necessary and sufficient to account for the 144 symmetries of the 6-j coefficient.

Using the properties satisfied by the elements of the 4×3 Bargmann–Shelepin symbol, the numerator and denominator variables belonging to the set II of ${}_4F_3(1)$s, given by (6.47), can be shown to be

$$\begin{gathered} A' = R_{q2} + R_{r1} + R_{s3} + 2, \quad B' = -R_{p1}, \quad C' = -R_{p2}, \\ D = -R_{p3}, \quad E' = R_{q1} - R_{p1} + 1, \quad F' = R_{r1} - R_{p1} + 1, \\ G' = R_{s1} - R_{p1} + 1, \end{gathered} \tag{6.66}$$

for $(pqrs) = (1234)$ cyclically. The 4×3 symbol for this set II of ${}_4F_3(1)$s has been shown (Srinivasa Rao and Rajeswari 1989) to be

$$\begin{Bmatrix} a & b & e \\ d & c & f \end{Bmatrix} = \left\| \begin{matrix} -B' & -C' & -D' \\ E' - B' - 1 & E' - C' - 1 & E' - D' - 1 \\ F' - B' - 1 & F' - C' - 1 & F' - D' - 1 \\ G' - B' - 1 & G' - C' - 1 & G' - D' - 1 \end{matrix} \right\| \tag{6.67}$$

where the positive numerator variable A' does not appear in (6.67) and this non-standard 4×3 symbol exhibits 36 symmetries due the three column permutations and three row permutations. Since there exists a set II of numerator and denominator variables of the ${}_4F_3(1)$s due to the cyclic permutations of $(pqrs) = (1234)$ in (6.66), it follows that there exists a set II of four non-standard 4×3 symbols equivalent to the standard Bargmann–Shelepin symbol (6.60).

6.7 The set of Bailey transformations

In 1940, in a note, Hardy (1940, p 111) stated that

$$\frac{1}{\Gamma(s, \beta_1, \beta_2)} {}_3F_2(\alpha_1, \alpha_2, \alpha_3; \beta_1, \beta_2; 1) \tag{6.68}$$

is a symmetric function of five arguments: $s + \alpha_1$, $s + \alpha_2$, $s + \alpha_3$, β_1, β_2, where $s = \beta_1 - \beta_2 - \alpha_1 - \alpha_2 - \alpha_3$—called the parameter excess—is a parameter and consequence of the Thomae transformation:

$$\begin{aligned} {}_3F_2(a, b, c; d, e; 1) = & \frac{\Gamma(d, e, s)}{\Gamma(a, s + b, s + c)} \\ & \times {}_3F_2(d - a, e - a, s; s + b, s + c; 1). \end{aligned} \tag{6.69}$$

Accordingly, that the function

$$f(x_1, x_2, x_3, x_4, x_5) = \frac{1}{\Gamma(s, 2x_4, 2x_5)} \times {}_3F_2(2x_1 - s, 2x_2 - s, 2x_3 - s; 2x_4, 2x_5; 1) \tag{6.70}$$

is symmetric in all the five variables is a consequence of (6.69) can be proved as follows: the function

$$f(\vec{x}) \equiv f(x_1, x_2, x_3, x_4, x_5) \tag{6.71}$$

is manifestly invariant for the permutations of (x_1, x_2, x_3) and (x_4, x_5), as these are the numerator and denominator variables of the ${}_3F_2(1)$. Consider now the permutation

$$p\colon\ x_1 \to x_2 \to x_3 \to x_4 \to x_5, \tag{6.72}$$

which is a permutation of order five. Upon relabeling the parameters of the ${}_3F_2(1)$ in (6.70) as ${}_3F_2(a, b, c; d, e; 1)$, it is straightforward to see that corresponding to

$$f(\vec{x}) = f(\vec{p}\cdot\vec{x}) \tag{6.73}$$

we get

$${}_3F_2(a, b, c; d, e; 1) = \frac{\Gamma(d, s)}{\Gamma(d - a, s + a)} {}_3F_2(a, e - b, e - c; e, s + a; 1), \tag{6.74}$$

on which are superposed the permutations of the numerator and denominator variables. Since $f(\vec{x})$ is invariant under this permutation p of order five, and under the transposition $x_4 \to x_5$ manifestly, the group generated by these two generators (viz p and the transposition) is the complete group of permutations on five elements. That is the symmetric group S_5 (see, for instance Budden 1972).

The two-term relation for the terminating Saalschützian

$${}_4F_3(A, B, C, -n; E, F, G; 1)$$

is given by Bailey (1935) as

$${}_4F_3(A, B, C, D, -n; E, F, G; 1) = \frac{(F - G)_n\,(G - C)_n}{(F)_n\,(G)_n} \times {}_4F_3(E - A, E - B, C, -n; E, E + F - A - B, E + G - A - B; 1), \tag{6.75}$$

where $(a)_n$ is the Pochammer symbol defined for $n \geqslant 1$. As in the case of the terminating ${}_3F_2(1)$ transformations, a recursive use of the above terminating

Saalschützian ${}_4F_3(1)$ results in the following set of 20 transformations, (Srinivasa Rao *et al* 1998).

$$
\begin{aligned}
{}_4F_3\left(\begin{matrix}A,\ B,\ C,\ -n\\ E,\ F,\ G\end{matrix}\right) &\equiv {}_4F_3(A,\ B,\ C,\ -n;\ E,\ F,\ G;\ 1)\\
&= \frac{(F-C)_n(G-C)_n}{(F)_n(G)_n}\,{}_4F_3\left(\begin{matrix}E-A,\ E-B,\ C,\ -n\\ E,\ E+F-A-B,\ E+G-A-B\end{matrix}\right)
\end{aligned} \tag{6.76}
$$

$$
= {}_4F_3\left(\begin{matrix}A,\ B,\ C,\ -n\\ E,\ F,\ G\end{matrix}\right)(\text{identity}) \tag{6.77}
$$

$$
= \frac{(E-C)_n(G-C)_n}{(E)_n(G)_n}\,{}_4F_3\left(\begin{matrix}F-A,\ F-B,\ C,\ -n\\ E+F-A-B,\ F,\ E+G-A-B\end{matrix}\right) \tag{6.78}
$$

$$
= \frac{(E-C)_n(F-C)_n}{(E)_n(F)_n}\,{}_4F_3\left(\begin{matrix}G-A,\ G-B,\ C,\ -n\\ E+G-A-B,\ F+G-A-B,\ G\end{matrix}\right) \tag{6.79}
$$

$$
= \frac{(F-B)_n(G-B)_n}{(F)_n(G)_n}\,{}_4F_3\left(\begin{matrix}E-A,\ B,\ E-C,\ -n\\ E,\ E+F-A-C,\ E+G-A-C\end{matrix}\right) \tag{6.80}
$$

$$
\begin{aligned}
&= (-1^n)\frac{(A)_n(G-B)_n(G-C)_n}{(E)_n(F)_n(G)_n}\\
&\times {}_4F_3\left(\begin{matrix}E-A,\ F-A,\ E+F-A-B-C,\ -n\\ E+F-A-B,\ E+F-A-C,\ E+F+G-2A-B-C\end{matrix}\right)
\end{aligned} \tag{6.81}
$$

$$
\begin{aligned}
&= (-1)^n\frac{(A)_n(F-B)_n(F-C)_n}{(E)_n(F)_n(G)_n}\\
&\times {}_4F_3\left(\begin{matrix}E-A,\ G-A,\ E+G-A-B-C,\ -n\\ E+G-A-B,\ E+G-A-C,\ E+F+G-2A-B-C\end{matrix}\right)
\end{aligned} \tag{6.82}
$$

$$
= \frac{(F-A)_n(G-A)_n}{(F)_n(G)_n}\,{}_4F_3\left(\begin{matrix}A,\ E-B,\ E-C,\ -n\\ E,\ E+F-B-C,\ E+G-B-C\end{matrix}\right) \tag{6.83}
$$

$$= (-1)^n \frac{(G-A)_n(B)_n(G-C)_n}{(E)_n(F)_n(G)_n} \times {}_4F_3\left(\begin{matrix} E-B,\ F-B,\ E+F-A-B-C,\ -n \\ E+F-A-B,\ E+F-A-C,\ E+F+G-A-2B-C \end{matrix}\right) \tag{6.84}$$

$$= (-1)^n \frac{(F-A)_n(B)_n(F-C)_n}{(E)_n(F)_n(G)_n} \times {}_4F_3\left(\begin{matrix} E-B,\ G-B,\ E+G-A-B-C,\ -n \\ E+G-A-B,\ E+G-B-C,\ E+F+G-A-2B-C \end{matrix}\right) \tag{6.85}$$

$$= \frac{(E-A)_n(G-A)_n}{(E)_n(G)_n}\ {}_4F_3\left(\begin{matrix} A,\ F-B,\ F-C,\ -n \\ E+F-B-C,\ F,\ F+G-B-C \end{matrix}\right) \tag{6.86}$$

$${}_4F_3\left(\begin{matrix} A,\ B,\ C,\ -n \\ E,\ F,\ G \end{matrix}\right) = (-1)^n \frac{(A)_n(E-B)_n(E-C)_n}{(E)_n(F)_n(G)_n} \times {}_4F_3\left(\begin{matrix} F-A,\ G-A,\ F+G-A-B-C,\ -n \\ F+G-A-B,\ F+G-A-C,\ E+F+G-2A-B-C \end{matrix}\right) \tag{6.87}$$

$$= \frac{(E-B)_n(G-B)_n}{(E)_n(G)_n}\ {}_4F_3\left(\begin{matrix} F-A,\ B,\ F-C,\ -n \\ E+F-A-C,\ F,\ F+G-A-C \end{matrix}\right) \tag{6.88}$$

$$= \frac{(E-A)_n(F-A)_n}{(E)_n(F)_n}\ {}_4F_3\left(\begin{matrix} A,\ G-B,\ G-C,\ -n \\ E+G-B-C,\ F+G-B-C,\ G \end{matrix}\right) \tag{6.89}$$

$$= \frac{(E-B)_n(F-B)_n}{(E)_n(F)_n}\ {}_4F_3\left(\begin{matrix} G-A,\ B,\ G-C,\ -n \\ E+G-A-C,\ F+G-A-C,\ G \end{matrix}\right) \tag{6.90}$$

$$= (-1)^n \frac{(G-A)_n(G-B)_n(C)_n}{(E)_n(F)_n(G)_n} \times {}_4F_3\left(\begin{matrix} E-C,\ F-C,\ E+F-A-B-C,\ -n \\ E+F-A-C,\ E+F-B-C,\ E+F+G-A-B-2C \end{matrix}\right) \tag{6.91}$$

$$= (-1)^n \frac{(E-A)_n (B)_n (E-C)_n}{(E)_n (F)_n (G)_n} \times {}_4F_3\left(\begin{matrix} F-B,\ G-B,\ F+G-A-B-C,\ -n \\ F+G-A-B,\ F+G-B-C,\ E+F+G-A-2B-C \end{matrix}\right) \tag{6.92}$$

$$= (-1)^n \frac{(F-A)_n (F-B)_n (C)_n}{(E)_n (F)_n (G)_n} \times {}_4F_3\left(\begin{matrix} E-C,\ G-C,\ E+G-A-B-C,\ -n \\ E+G-A-C,\ E+G-B-C,\ E+F+G-A-B-2C \end{matrix}\right) \tag{6.93}$$

$$= (-1)^n \frac{(E-A)_n (E-B)_n (C)_n}{(E)_n (F)_n (G)_n} \times {}_4F_3\left(\begin{matrix} F-C,\ G-C,\ F+G-A-B-C,\ -n \\ F+G-A-C,\ F+G-B-C,\ E+F+G-A-B-2C \end{matrix}\right) \tag{6.94}$$

$${}_4F_3\left(\begin{matrix} A,\ B,\ C,\ -n \\ E,\ F,\ G \end{matrix}\right) = (-1)^n \frac{(A)_n (B)_n (C)_n}{(E)_n (F)_n (G)_n} \times {}_4F_3\left(\begin{matrix} \mathrm{E+F-A-B-C,\ F+G-A-B-C,\ E+G-A-B-C,} \\ -n \\ \mathrm{E+F+G-2A-B-C,\ E+F+G-A-2B-C,\ E+F+G} \\ \mathrm{-A-B-2C} \end{matrix}\right). \tag{6.95}$$

We construct the function

$$\begin{aligned} & f(x_1, x_2, x_3, x_4, x_5, x_6) \\ & = (x_1 + x_2 + x_3 + x_4,\ x_1 + x_2 + x_3 + x_5,\ x + 1 + x_2 + x_3 + x_6)_n \\ & \times {}_4F_3\left(\begin{matrix} x_1 + x_2,\ x_2 + x_3,\ x_1 + x_3,\ -n \\ x_1 + x_2 + x_3 + x_4,\ x_1 + x_2 + x_3 + x_5,\ x + 1 + x_2 + x_3 + x_6 \end{matrix}\right) \end{aligned} \tag{6.96}$$

with $(x, y, \ldots)_n = (x)_n (y)_n \cdots$ and

$$x_1 + x_2 + x_3 + x_4 + x_5 + x_6 = 1 - n, \tag{6.97}$$

for the negative integer n, which, being symmetric in all six variables, is a consequence of the Saalschützian condition on the numerator and denominator variables of the ${}_4F_3(1)$. We note that the given function is obviously invariant under the permutations of x_1, x_2, x_3 and x_4, x_5, x_6. Consider the cyclic permutation of order 6,

$$p: x_1 \rightarrow x_6 \rightarrow x_4 \rightarrow x_2 \rightarrow x_5 \rightarrow x_3 \rightarrow x_1. \tag{6.98}$$

We relabel the parameters of ${}_4F_3(1)$ in $f(\vec{x})$ (6.96) as: ${}_4F_3(A, B, C, -n; E, F, G)$. It is straightforward to see that

$$f(\vec{x}) = f(p\,\vec{x}). \tag{6.99}$$

We get 20 ${}_4F_3(1)$ transformations (6.76) to (6.95) on which are superposed the three-numerator variable permutations of (A, B, C) and the three-denominator variable permutations of (E, F, G), giving rise to a total of $20 \times 6 \times 6 = 720$ transformations, which constitute the representations for the elements of the symmetric group S_6. In other words, $f(\vec{x})$ is invariant under the permutation p of order 6. This has been shown by Beyer *et al* (1987). Here, following Srinivasa Rao *et al* (1998), we establish that the result is simply, *à la* Hardy (1940), a succinct, quintessential one line statement.

6.8 Basis states and symmetries of the 6-*j* coefficient

The classical tetrahedral symmetries of the 6-*j* coefficient arise due to the three column permutations of the three columns in the 6-*j* symbol, and the interchange of any pair of elements in one row with the corresponding elements in the 6-*j* symbol. The highly symmetric form of the 6-*j* coefficient, first noted by Regge (1958), clearly exhibits the 144 symmetries that arise due to the four permutations of the four αs and the three permutations of the three βs. These *mixed* symmetries are made of the six Regge symmetries given by the identity and (6.19)–(6.23) on which are superposed the 24 *classical* tetrahedral symmetries.

Let $\mathcal{R}_x^6$ be the six dimensional angular momentum space and let the column (ket) vector in this space be

$$|\vec{x}\rangle \equiv |a, b, c, d, e, f\rangle \,\varepsilon\, \mathcal{R}_x^6. \tag{6.100}$$

The *classical* symmetry generators g_i^x, $i = 1, 2, 3$, in $\mathcal{R}_x^6$ are

$$\begin{pmatrix} 0&1&0&0&0&0\\ 1&0&0&0&0&0\\ 0&0&0&1&0&0\\ 0&0&1&0&0&0\\ 0&0&0&0&1&0\\ 0&0&0&0&0&1 \end{pmatrix}, \begin{pmatrix} 1&0&0&0&0&0\\ 1&0&0&0&1&0\\ 0&0&0&1&0&1\\ 0&0&0&1&0&0\\ 0&1&0&0&0&0\\ 0&0&1&0&0&1 \end{pmatrix}, \tag{6.101}$$
$$\begin{pmatrix} 0&0&0&1&0&0\\ 0&0&1&0&0&0\\ 0&1&0&0&0&0\\ 1&0&0&0&0&0\\ 0&0&0&0&1&0\\ 0&0&0&0&0&1 \end{pmatrix}.$$

It is straightforward to write down the generators γ_i, $i = 1, 2, 3, 4$, for the $S_4 \times S_3$ as

$$\begin{pmatrix} 0&1&0&0&0&0&0\\ 0&0&1&0&0&0&0\\ 0&0&0&1&0&0&0\\ 1&0&0&0&0&0&0\\ 0&0&0&0&1&0&0\\ 0&0&0&0&0&1&0\\ 0&0&0&0&0&0&1 \end{pmatrix}, \begin{pmatrix} 0&1&0&0&0&0&0\\ 1&0&0&0&0&0&0\\ 0&0&1&0&0&0&0\\ 0&0&0&1&0&0&0\\ 0&0&0&0&1&0&0\\ 0&0&1&0&0&1&0\\ 0&0&0&0&0&0&1 \end{pmatrix},$$
$$\begin{pmatrix} 1&0&0&0&0&0&0\\ 0&1&0&0&0&0&0\\ 0&0&1&0&0&0&0\\ 0&0&0&1&0&0&0\\ 0&0&0&0&0&1&0\\ 0&0&0&0&0&0&1\\ 0&0&0&0&1&0&0 \end{pmatrix}, \begin{pmatrix} 1&0&0&0&0&0&0\\ 0&1&0&0&0&0&0\\ 0&0&1&0&0&0&0\\ 0&0&0&1&0&0&0\\ 0&0&0&0&0&1&0\\ 0&0&0&0&1&0&0\\ 0&0&0&0&0&0&1 \end{pmatrix}. \tag{6.102}$$

These are projected onto the invariant subspace, using the projection matrix

$$\mathcal{P} = I_{7\times 7} - \frac{1}{7}\hat{P} \tag{6.103}$$

where $I_{7\times 7}$ is the 7-dimensional identity matrix and

$$\hat{P} = \begin{pmatrix} 1&1&1&1&-1&-1&-1\\ 1&1&1&1&-1&-1&-1\\ 1&1&1&1&-1&-1&-1\\ 1&1&1&1&-1&-1&-1\\ -1&-1&-1&-1&1&1&1\\ -1&-1&-1&-1&1&1&1\\ -1&-1&-1&-1&1&1&1 \end{pmatrix}. \tag{6.104}$$

Let T_3 be the transformation from $\mathcal{R}_x^6$ to $\mathcal{R}_y^7$:

$$T_3\colon\ \mathcal{R}_x^6 \to \mathcal{R}_y^7 \tag{6.105}$$

and let T_4 be the reverse of the above transformation from $\mathcal{R}_y^7$ to $\mathcal{R}_x^6$:

$$T_4\colon\ \mathcal{R}_y^7 \to \mathcal{R}_x^6. \tag{6.106}$$

Explicitly, T_3 and T_4 are given by

$$T_3 = \begin{pmatrix} 1&1&0&0&1&0\\ 0&0&1&1&1&0\\ 1&0&1&0&0&1\\ 0&1&0&1&0&1\\ 1&1&1&1&0&0\\ 0&1&1&0&1&1 \end{pmatrix}, \frac{1}{2}\begin{pmatrix} 1&0&1&0&0&0&-1\\ 1&0&0&1&0&-1&0\\ 0&1&1&0&0&-1&0\\ 0&1&0&1&0&0&-1\\ 1&1&0&0&-1&0&0\\ 0&1&1&1&-1&0&0 \end{pmatrix}\cdot\mathcal{P}, \tag{6.107}$$

such that the identity is

$$\begin{aligned} T_3 \,.\, T_4 &= I, \quad \text{in} \quad \mathcal{R}^6_x, \\ T_4 \,.\, T_3 &= I, \quad \text{in} \quad \mathcal{R}^7_y, \end{aligned} \tag{6.108}$$

and we note that the former is an identity only after the condition

$$\sum_{i=1}^{4} \alpha_i = \sum_{j=1}^{3} \beta_j \tag{6.109}$$

is satisfied.

From the generators g_i^x (6.36) in $\mathcal{R}^6_x$, we obtain the corresponding generators g_i^y in $\mathcal{R}^7_y$ through

$$g_i^y = T_3 \,.\, g_i^x \,.\, T_4. \tag{6.110}$$

Conversely, starting with the five generators (6.102)—the fifth being the $I_{5\times 5}$ identity matrix—in $\mathcal{R}^7_y$, the corresponding generators γ_i^x in $\mathcal{R}^6_x$ can be obtained through

$$\gamma_i^x = T_4 \,.\, \gamma_i^y \,.\, T_3. \tag{6.111}$$

These 6 × 6 matrices are directly related to the mixed symmetries of the 6-j coefficient when they operate on the basis vector $|\vec{x}\rangle$. By mixed symmetries we mean the Regge symmetries on which the *classical* tetrahedrral symmetries are superposed. In the next section we will explicate these generators g_i^y and γ_i^x, and show how the latter generate the 144 symmetries of the 6-j coefficient.

6.9 Generators of the symmetries of the 6-*j* coefficient

These generators are in terms of the αs and the βs given in (6.11), and so will be denoted as $g_i^{\alpha,\beta}$ in the basis space $\mathcal{R}^5_{\alpha,\beta}$ as:

$$\begin{pmatrix} 0 & 0 & 1 & -1 & 0 \\ 0 & 0 & 1 & 0 & -1 \\ 0 & 0 & 1 & 0 & 0 \\ -1 & 0 & 1 & 0 & 0 \\ 0 & -1 & 1 & 0 & 0 \end{pmatrix}, \begin{pmatrix} 1 & 0 & 0 & 0 & 0 \\ 1 & 0 & 0 & -1 & 0 \\ 1 & 0 & 0 & 0 & -1 \\ 1 & -1 & 0 & 0 & 0 \\ 1 & 1 & -1 & 0 & 0 \end{pmatrix}, \begin{pmatrix} 0 & 0 & 1 & -1 & 0 \\ 0 & 0 & 1 & 0 & 0 \\ 0 & 0 & 1 & 0 & -1 \\ 0 & -1 & 1 & 0 & 0 \\ -1 & 0 & 1 & 0 & 0 \end{pmatrix}. \tag{6.112}$$

The generators $\gamma_i^{j,m}$ in the basis space $\mathcal{R}^6_{j,m}$ are

$$\frac{1}{2}\begin{pmatrix} 1 & 1 & 0 & \frac{1}{3} & \frac{1}{3} & -\frac{2}{3} \\ 1 & 1 & 0 & -\frac{1}{3} & -\frac{1}{3} & \frac{2}{3} \\ 0 & 0 & 2 & 0 & 0 & 0 \\ 1 & -1 & 0 & 1 & -1 & 0 \\ 1 & -1 & 0 & -1 & 1 & 0 \\ -2 & 2 & 0 & 0 & 0 & 0 \end{pmatrix}, \frac{1}{2}\begin{pmatrix} 1 & 0 & 1 & -\frac{1}{3} & \frac{2}{3} & -\frac{1}{3} \\ 0 & 2 & 0 & 0 & 0 & 0 \\ 1 & 0 & 1 & \frac{1}{3} & -\frac{2}{3} & \frac{1}{3} \\ -1 & 0 & 1 & 1 & 0 & -1 \\ 2 & 0 & -2 & 0 & 0 & 0 \\ -1 & 0 & 1 & -1 & 0 & 1 \end{pmatrix},$$
$$\frac{1}{2}\begin{pmatrix} 1 & 1 & 0 & -\frac{1}{3} & -\frac{1}{3} & \frac{2}{3} \\ 1 & 1 & 0 & \frac{1}{3} & \frac{1}{3} & -\frac{2}{3} \\ 0 & 0 & 2 & 0 & 0 & 0 \\ -1 & 1 & 0 & 1 & -1 & 0 \\ -1 & 1 & 0 & -1 & 1 & 0 \\ 2 & -2 & 0 & 0 & 0 & 0 \end{pmatrix}. \tag{6.113}$$

If we denote the set of matrix generators $g_i^{\alpha,\beta}$ by a, b, c, then they satisfy the properties

$$\begin{gathered} a^2 = b^2 = c^2 = I, \quad bc = cb, \\ ac = ca, \quad ba = (ab)^2 \to (ab)^3 = 1. \end{gathered} \tag{6.114}$$

The only independent elements that can be generated by forming all possible products of all possible powers of a, b, c are

$$I,\ a,\ b,\ c,\ ab,\ bc,\ ca,\ ba,\ abc,\ bac,\ aba,\ babc, \tag{6.115}$$

where I is the five dimensional unit matrix. The multiplication table for the elements reveals that they form a group with the following subgroups

$$\begin{gathered} \{1, a\},\ \{a, b\},\ \{1, c\},\ \{1, a, c, ac\}, \\ \{a, b, c, bc\},\ \{a, a, b, ab, ba, aba\}, \end{gathered} \tag{6.116}$$

with $\{1, c\}$ being the center of the group, since c commutes with every element of the group.

In the case of the 6-j coefficient, the generators g_i^y in $\mathcal{R}^7_y$, after the condition (6.109) is taken into account, are

$$\begin{pmatrix} 1&0&0&0&0&0&0\\ 0&1&0&0&0&0&0\\ 0&0&0&1&0&0&0\\ 0&0&1&0&0&0&0\\ 0&0&0&0&1&0&0\\ 0&0&0&0&0&0&1\\ 0&0&0&0&0&1&0 \end{pmatrix}, \begin{pmatrix} 1&0&0&0&0&0&0\\ 0&0&0&1&0&0&0\\ 0&0&1&0&0&0&0\\ 0&1&0&0&0&0&0\\ 0&0&0&0&0&1&0\\ 0&0&0&0&1&0&0\\ 0&0&0&0&0&0&1 \end{pmatrix}, \begin{pmatrix} 0&1&0&0&0&0&0\\ 1&0&0&0&0&0&0\\ 0&0&0&1&0&0&0\\ 0&0&1&0&0&0&0\\ 0&0&0&0&1&0&0\\ 0&0&0&0&0&1&0\\ 0&0&0&0&0&0&1 \end{pmatrix}. \tag{6.117}$$

The generators γ_i^x in $\mathcal{R}_x^6$ obtained are

$$\frac{1}{2}\begin{pmatrix} 0&0&0&2&0&0\\ 0&1&1&0&1&-1\\ 0&1&1&0&-1&1\\ 2&0&0&0&0&0\\ 0&-1&1&0&1&1\\ 1&-1&0&0&1&1 \end{pmatrix}, \frac{1}{2}\begin{pmatrix} 1&-1&1&1&0&0\\ -1&1&1&1&0&0\\ 1&1&1&-1&0&0\\ 1&1&-1&1&0&0\\ 0&0&0&0&2&0\\ 0&0&0&0&0&2 \end{pmatrix},$$
$$\frac{1}{2}\begin{pmatrix} 1&0&0&-1&1&1\\ 1&1&-1&1&0&0\\ 1&-1&1&1&0&0\\ -1&0&0&1&1&1\\ 0&1&1&0&1&-1\\ 0&1&1&0&-1&1 \end{pmatrix}, \frac{1}{2}\begin{pmatrix} 2&0&0&0&0&0\\ 0&1&-1&0&1&1\\ 0&-1&1&0&1&1\\ 0&0&0&2&0&0\\ 0&1&1&0&1&-1\\ 0&1&1&0&-1&1 \end{pmatrix}. \tag{6.118}$$

These latter 6×6 matrices are directly related to the Regge symmetries (1958) of the 6-j coefficient when they operate on the basis vector $|a, b, c, d, e, f\rangle$ and they are

$$\frac{1}{2}\begin{pmatrix} 0&0&0&2&0&0\\ 0&1&1&0&1&-1\\ 0&1&1&0&-1&1\\ 2&0&0&0&0&0\\ 0&-1&1&0&1&1\\ 1&-1&0&0&1&1 \end{pmatrix}\begin{pmatrix} a\\ b\\ c\\ d\\ e\\ f \end{pmatrix} = \begin{Bmatrix} a' & b' & e'\\ d' & c' & f' \end{Bmatrix}$$
$$= \begin{Bmatrix} d & \frac{1}{2}(b+c+e-f) & \frac{1}{2}(-b+c+e+f)\\ a & \frac{1}{2}(b+c-e+f) & \frac{1}{2}(b-c+e+f) \end{Bmatrix} \tag{6.119}$$

$$\frac{1}{2}\begin{pmatrix} 1 & -1 & 1 & 1 & 0 & 0 \\ -1 & 1 & 1 & 1 & 0 & 0 \\ 1 & 1 & 1 & -1 & 0 & 0 \\ 1 & 1 & -1 & 1 & 0 & 0 \\ 0 & 0 & 0 & 0 & 2 & 0 \\ 0 & 0 & 0 & 0 & 0 & 2 \end{pmatrix}\begin{pmatrix} a \\ b \\ c \\ d \\ e \\ f \end{pmatrix} = \begin{Bmatrix} a' & b' & e' \\ d' & c' & f' \end{Bmatrix}$$
$$= \begin{Bmatrix} \frac{1}{2}(a-b+c+d) & \frac{1}{2}(-a+b+c+d) & e \\ \frac{1}{2}(a+b-c+d) & \frac{1}{2}(a+b+c-d) & f \end{Bmatrix} \tag{6.120}$$

$$\frac{1}{2}\begin{pmatrix} 1 & 0 & 0 & -1 & 1 & 1 \\ 1 & 1 & -1 & 1 & 0 & 0 \\ 1 & -1 & 1 & 1 & 0 & 0 \\ -1 & 0 & 0 & 1 & 1 & 1 \\ 0 & 1 & 1 & 0 & 1 & -1 \\ 0 & 1 & 1 & 0 & -1 & 1 \end{pmatrix}\begin{pmatrix} a \\ b \\ c \\ d \\ e \\ f \end{pmatrix} = \begin{Bmatrix} a' & b' & e' \\ d' & c' & f' \end{Bmatrix}$$
$$= \begin{Bmatrix} \frac{1}{2}(a-d+e+f) & \frac{1}{2}(a+b-c+d) & \frac{1}{2}(b+c+e-f) \\ \frac{1}{2}(-a+d+e+f) & \frac{1}{2}(a-b+c+d) & \frac{1}{2}(b+c-e+f) \end{Bmatrix} \tag{6.121}$$

$$\frac{1}{2}\begin{pmatrix} 2 & 0 & 0 & 0 & 0 & 0 \\ 0 & 1 & -1 & 0 & 1 & 1 \\ 0 & -1 & 1 & 0 & 1 & 1 \\ 0 & 0 & 0 & 2 & 0 & 0 \\ 0 & 1 & 1 & 0 & 1 & -1 \\ 0 & 1 & 1 & 0 & -1 & 1 \end{pmatrix} = \begin{Bmatrix} a' & b' & e' \\ d' & c' & f' \end{Bmatrix}$$
$$= \begin{Bmatrix} a & \frac{1}{2}(b-c+e+f) & \frac{1}{2}(b+c+e-f) \\ d & \frac{1}{2}(-b+c+e+f) & \frac{1}{2}(b+c-e+f) \end{Bmatrix}. \tag{6.122}$$

These are *mixed* symmetries in the sense that they are the Regge symmetries of the 6-j coefficient on which are superposed the *classical tetrahedral* symmetries.

6.10 Summary and remarks

The quintessential and seminal remark of Hardy in his Notes on Lecture VII in *Ramanujan: Twelve Lectures on Subjects Suggested by His Life and Work* (1940) is the following: formula (7.3.3) is equivalent to (1), of section 3.2, of Bailey's tract. It is an expression of the theorem that

$$\frac{1}{\Gamma(\beta_1)\Gamma(\beta_2)\Gamma(\beta_1+\beta_2-\alpha_1-\alpha_2-\alpha_3)} F\begin{pmatrix}\alpha_1, \alpha_2, \alpha_3\\ \beta_1, \beta_2\end{pmatrix} \tag{6.123}$$

is a symmetric function of the five arguments

$$\beta_1, \beta_2, \beta_1+\beta_2-\alpha_2-\alpha_3, \beta_1+\beta_2-\alpha_3-\alpha_1,$$
$$\beta_1+\beta_2-\alpha_1-\alpha_2.$$

This clearly implies a group theoretical interpretation for formula (7.3.3), which is a ${}_3F_2(1)$ transformation of Thomae (1879):

$$\begin{aligned}&\frac{\Gamma(x+y+s+1)}{\Gamma(x+s+1)\Gamma(y+s+1)} F\begin{pmatrix}-a, -b, x+y+s+1\\ x+s+1, y+s+1\end{pmatrix}\\ &=\frac{\Gamma(a+b+s+1)}{\Gamma(a+s+1)\Gamma(b+s+1)} F\begin{pmatrix}-x, -y, a+b+s+1\\ a+s+1, b+s+1\end{pmatrix}.\end{aligned} \tag{6.124}$$

In 1987, Beyer, Louck, and Stein rediscovered that Thomae's identity between two ${}_3F_2(1)$ hypergeometric series' of unit arguments—together with the trivial invariance under separate permutations of the numerator and denominator variables—implies that the symmetric group S_5 is an invariance group of this series. They also showed that Bailey's identity for the Saalschützian ${}_4F_3(1)$ series has S_6 as its invariance group. Srinivasa Rao *et al* (1992) studied the group theory of the terminating ${}_3F_2(1)$ series—a case not studied by Beyer, Louck, and Stein (1987)—and found all the invariant subgroups of the 72-element group of the ${}_3F_2(1)$ transformations, which we detail in chapter 5.

References

Bailey W N 1935 *Generalized Hypergeometric Series* Cambridge Tracts in Mathematics and Mathematical Physics 32 (Cambridge: Cambridge University Press)

Bandzaitis A A, Karosene A V, Sauunikas Y U and Yustis A P 1964 *Dokl. Akad. Nauka SSR* **154** 812

Bargman V 1962 *Rev. Mod. Phys.* **34** 829

Beyer W A, Louck J D and Stein P R 1987 *J. Math. Phys.* **28** 497

Biedenharn L C and Louck J D 1981 *Angular Momentum in Quantum Physics: Theory and Application* (Encyclopedia of Mathematics and its Applications vol 8) (Cambridge: Cambridge University Press)

Biedenharn L C and Louck J D 1984 *The Racah Wigner Algebra in Quantum Theory* (Encyclopedia of Mathematics and its Applications vol 9) (Cambridge: Cambridge University Press)

Budden F J 1972 *The Fascination of Groups* (Cambridge: Cambridge University Press)

D'Adda A, D'Auria R and Ponzano G 1971 On generalized Wigner and Racah coefficients *Internal Report* September 1971. Also, see 1972 *Proc. of the Int. School of Physics 'Enrico Fermi'* vol 54 and 1972 Lettre al Nuovo Cimento, 5, 973

D'Adda A, D'Auria R and Ponzano G 1974 *Nuovo Cimentro* **23** 69
Hardy G H 1940 *Ramanujan: Twelve Lectures Inspired by his Life and Work* (Cambridge: Cambridge University Press)
Holman W J and Biedenbarn L C 1968 *Ann. Phys.* **47** 205
Holman W J and Biedenbarn L C 1966 *Ann. Phys.* **39** 1
Louck J D, Beyer W A, Biedernharn L C and Stein P R 1986 Symmetries of some hypergeometric series: implications for 3-j and 6-j symbols *Los Alamos National Laboratory report* LA-UR-86-4203
Minton B M 1970 *Math. Phys.* **11** 3061
Racah G 1942a *Phys. Rev.* **61** 186
Racah G 1942b *Phys. Rev.* **62** 438
Racah G 1943 *Phys. Rev.* **63** 367
Racah G 1949 *Phys. Rev.* **76** 1352
Regge T 1958 *Nuovo Cimento* **10** 544
Regge T 1959 *Nuovo Cimento* **11** 116
Regge T 1960 *Nuovo Cimento* **18** 947–56
Rose M E 1955 *Multipole Fields* (New York: Wiley)
Shelepin L A 1964 *Sov. Phys. JETP* **19** 702
Slater L J 1966 *Generalized Hypergeometric Functions* (Cambridge: Cambridge University Press)
Smorodinskii Ya A and Shelepin Leonid A 1972 *Sov. Phys. Uspekhi* **15** 1–24
Srinivasa Rao K, Santhanam T S and Venkatesh K 1975 *J. Math. Phys.* **16** 1528
Srinivasa Rao K and Venkatesh K 1978 *Comput. Phys. Commun.* **15** 227
Srinivasa Rao K 1978 *J. Phys.* **11A** L69
Srinivasa Rao K and Rajeswari V 1985 *Rev. Mex. Fis.* **31** 575
Srinivasa Rao K and Rajeswari V 1989 *J. Math. Phys.* **30** 1016
Srinivasa Rao K, Van der Jeugt J, Raynal J, Jagannathan R and Rajeswari V 1992 *J. Phys. A: Math. Gen.* **25** 861–76
Srinivasa Rao K, Doebner H -D and Natterman P 1998 *Generalized Hypergeometric Series and the Symmetries of 3-j and 6-j Coefficients Number Theoretic Methods: Developments in Mathematics* ed S Kanemitsu and C Jia vol 8 (New York: Springer)
Thomae J 1879 *J. für Math.* **87** 26–73
Varshalovich D A, Moskalev A N and Khersonskii V K 1975 *A Quantum Theory of Angular Momentum* (Leningrad: Nauka) in Russian
Yakimiw E 1971 *J. Math. Phys.* **12** 1134
Yutsis A P and Bandzaitis A A 1977 *The Theory of Angular Momentum in Quantum Mechanics* (Vilnius: Mokslas) in Russian

Chapter 7

Double and triple hypergeometric series

7.1 Introduction: a history

In 1828, Clausen introduced a series—a generalization of the Gauss hypergeometric series (1.3)—by increasing the number of numerator and denominator variables from two and one to three and two, respectively, and defined this series as[1]

$$ {}_3F_2(a,\, b,\, c;\, d,\, e;\, z) = \sum_{n=1}^{\infty} \frac{(a)_n(b)_n(c)_n}{(d)_n(e)_n} \frac{z^n}{n!}. \tag{7.1} $$

In chapter 4, the group theory of terminating and non-terminating [4] ${}_3F_2(a, b, c; d, e; 1)$ transformations was presented. Chapters 5 and 6 were devoted to the quantum theory of angular momentum (QTAM), and the natural way in which the coupling (3-j) and recoupling (6-j) coefficients lead to sets of hypergeometric functions and their group theoretical consequences.

In this chapter we introduce further generalizations of the Gauss series in one variable, by Appell (1925), to the corresponding theory in two variables, as a product of two ${}_2F_1$s. The work of Appell and Kampé de Fériet (1926) provided the starting point for the study of a multiple hypergeometric series in later years. A generalization of the products of ${}_pF_q$s—especially in the hypergeometric series in variables two and three—are referred to as the double and triple hypergeometric series. Generalized hypergeometric series' occur normally in evaluation of integrals involving special functions.

The 9-j coefficient has been shown to be a product of six 3-j coefficients, by Wigner (1940); a sum of all six independent projection quantum numbers, and also a product of three 6-j coefficients, summed over a single index. He also pointed out that the symmetries of the 9-j coefficient are more easily obtained from the symmetries of the 3-j coefficient. While the 3-j coefficient has 72 symmetries and

[1] See Lathrop and Stemkosk (2007).

the 9-j coefficient has 144 symmetries, the 9-j coefficient also has 72 symmetries (and not more). The symmetry group of the 9-j coefficient is written as

$$\begin{Bmatrix} j_1 & j_2 & j_{12} \\ j_3 & j_4 & j_{34} \\ j_{13} & j_{24} & J \end{Bmatrix} \equiv \begin{Bmatrix} \ell_1 & \ell_2 & L \\ s_1 & s_2 & S \\ j_1 & j_2 & J \end{Bmatrix} \tag{7.2}$$

where $S_3 \times S_3 \times S_2$ is a product of three column and three row permutations, as well as a reflection of its diagonality (or a transposition, if the entries of the 9-j coefficient are treated as a 3×3 matrix).

Besides the 3-j and 6-j coefficients in the QTAM (Srinivasa Rao and Rajeswari 1993), the recoupling coefficient that arises, due to two different ways of coupling four angular momenta, results in the 9-j coefficient; which is vital in the theoretical studies of atomic and nuclear physics. In this chapter, we present an introduction to the 9-j coefficient and show that its simplest series form is a triple series, as seen in Alisauskas and Jucys (1971). This triple series has been related by Srinivasa Rao and Rajeswari (1989) to a triple hypergeometric series. They showed that the *new* generalized hypergeometric function in the three variables is a special case of the general triple hypergeometric series studied by Lauricella (1893), Saran (1954), and Srivasatava and Karlsson (1985).

Surprisingly, the triple sum series of Alisauskas–Jucys is highly asymmetric and does not exhibit any of the symmetries of the 9-j coefficient. This turns out to be nothing short of a blessing in disguise for Srinivasa Rao and Rajeswari (1999), who showed that a certain 9-j coefficient expressed as a triple sum series has only one term, while its symmetries have a varying number of terms; one of them has as many as 33 761 terms.

The identification of the triple sum series by Srinivasa Rao and Rajeswari (1989) responds to the remark of A C T Wu (1972), which states that *the naive conjecture that the 9-j coefficient might also belong to some hypergeometric* ${}_pF_q$ *form turns out to be false. The best that can be said in this regard is that the 9-j symbol is a folded product of either* ${}_3F_2$ *or* ${}_4F_3$ *functions.* A conjecture based on the fact that Wigner had expressed the 9-j coefficient as either a sum of the products of six 3-j coefficients, or as a sum over a product of 6-j coefficients; coupled with the fact that the 3-j coefficient can be expressed as ${}_3F_2(1)$, and the 6-j coefficient can be expressed as ${}_4F_3(1)$. Wu (1972, 1973) derived a new explicit expression for the 9-j coefficient as a sixfold sum, which he stated *may be regarded as the analog of Racah's formula for the Racah coefficient*, in 1972 and showed that it is not related to any class of generalized hypergeometric function.

This identification of the triple sum series for the 9-j coefficient as a special case of the triple hypergeometric series led to several important consequences:

- First, it is possible to write down this series as a folded form, and this enables one to show that the new algorithm it is based on is computationally faster and more efficient for the usual range of parameters of interest to atomic and nuclear physics.

- Second, it was possible to define for the first time the polynomial zeros for the 9-j coefficient and provide a complete solution to the problem of polynomial zeros of degree.
- Third, it was possible to obtain not only all the known summation and transformation formulas involving the 9-j coefficient, but also discover new ones by studying singly and doubly stretched 9-j coefficients.

7.2 Multiple hypergeometric series

The series (7.1) enabled the introduction of the generalized hypergeometric series:

$$ {}_qF_p\left(\begin{matrix} a_1, a_2, \ldots a_p \\ b_1, b_2, \ldots b_q \end{matrix}; z\right) = \sum_{k=0}^{\infty} \frac{(a_1)_k(a_2)_k\cdots(a_q)_k}{(b_1)_k(b_2)_k\cdots(b_p)_k}\frac{z^k}{k!}, \tag{7.3} $$

where p and q are integers in the applications to be presented later in this monograph, $q = p + 1$; with the extra numerator variable being a negative integer, $-n$, for terminating the generalized hypergeometric series. In the following chapters, the ${}_3F_2(1)$ form is shown to be related to the Clebsch–Gordan or 3-j coefficient, and the ${}_4F_3(1)$ form related to the Racah or 6-j coefficient. These angular momentum coupling and recoupling coefficients are essential in the studies of spectra, and reactions in the fields of atomic, nuclear, and molecular physics.

An obvious generalization of the second order ODE (1.84) of Gauss is

$$ \{\theta(\theta + b_1 - 1)\cdots(\theta + b_p - 1) - z(\theta + a_1)\cdots(\theta + a_{p+1})\}y \ 0, \tag{7.4} $$

where $\theta \equiv z(d/dz)$, and $a_1, a_2, \ldots, a_{p+1}, b_1, b_2, \ldots, b_p$ are parameters. It can be readily shown that (7.2) is a generalization of the ${}_2F_1(z)$ series, and hence called the generalized hypergeometric series. Though it is a generalization of the Gauss series (1.5), it is not sufficiently wide to cover a simple series of the type ${}_1F_1(a; c; z)$, known as the confluent hypergeometric function, which is a solution for the differential equation (1.92). This equation is called the Kummer equation, and it can be obtained from the Gauss second order ODE (1.29) if we replace z in the Gauss differential equation (1.83) with x/b and let $b \to \infty$. The irregular singularity at ∞ of Kummer's equation is formed by the *confluence* of two regular singularities at b and ∞ (see Slater 1960). Hence, Kummer's equation is called the *confluent hypergeometric equation* (see Lakshminarayanan and Varadharajan 2015; chapter 12).

The theory of the Gauss hypergeometric series in one variable has been generalized by Appell to the corresponding theory in two variables (see Appell and Kampé de Fériet 1926). This work provided the impetus for the study of the multiple hypergeometric series in later years. Consider the product of two Gaussian ${}_2F_1$ series each with a different variable:

$$ {}_2F_1(a, b; c; x)\,{}_2F_1(a', b'; c'; y) = \sum_{m=0}^{\infty} \frac{(a)_m(b)_m}{(c)_m}\frac{x^m}{m!} \times \sum_{n=0}^{\infty} \frac{(a)_n(b)_n}{(c)_n}\frac{x^n}{n!}. \tag{7.5} $$

This is a product of two single series', and not a genuine double-series as such. However, Appell replaced one or more of the three pairs of products, viz

$$(a)_m(a')_n,\ (b)_m(b')_n,\ (c)_m(c')_n, \tag{7.6}$$

by the corresponding products of the type

$$(a)_{m+n},\ (b)_{m+n},\ (c)_{m+n} \tag{7.7}$$

and arrived at five new functions. Of these, the one obtained by replacing all three pairs of products by their corresponding composites was shown by Slater (1966) to be an ordinary Gauss series in the one variable $x + y$:

$$_2F_1(a,\ b;\ c;\ x+y) = \sum_{m,n=0}^{\infty} \frac{(a)_{m+n}(b)_{m+n}}{(c)_{m+n}} \frac{x^m}{m!} \frac{y^n}{n!}. \tag{7.8}$$

Explicitly, the product of two Gauss $_2F_1$ series is

$$\begin{aligned} {}_2F_1(a,\ b;\ c;\ x){}_2F_1(a',\ b';\ c';\ y) &= \sum_{m=0}^{\infty} \frac{(a)_m(b)_m}{(c)_m} \frac{x^m}{m!} \sum_{n=0}^{\infty} \frac{(a)_n(b)_n}{(c)_n} \frac{y^n}{n!} \\ &\Rightarrow \sum_{m,n=0}^{\infty} \frac{(a)_{m+n}(b)_{m+n}}{(c)_{m+n}} \frac{x^m}{m!} \frac{y^n}{n!} \end{aligned} \tag{7.9}$$

$$_2F_1(a,\ b;\ c;\ x+y) = \sum_{n=0}^{\infty} \frac{(a)_n(b)_n}{(c)_n} \frac{(x+y)^n}{n!}. \tag{7.10}$$

Expanding the series on the right-hand side of both (7.9) and (7.10), gives rise to the same series

$$\begin{aligned} &1 + \frac{(a)_1(b)_1}{(c)_1}\frac{x}{1!} + \frac{(a)_1(b)_1}{(c)_1}\frac{y}{1!} + \frac{(a)_2(b)_0}{(c)_2}\frac{x^2}{2!} + \frac{(a)_2(b)_2}{(c)_2}\frac{xy}{1!1!} + \frac{(a)_2(b)_2}{(c)_2}\frac{y^2}{2!} \\ &+ \cdots \\ &= 1 + \frac{(a)_1(b)_1}{(c)_1}\frac{(x+y)}{1!} + \frac{(a)_2(b)_2}{(c)_2}\frac{(x+y)^2}{2!} + \cdots = {}_2F_1(a,\ b;\ c;\ x+y), \end{aligned} \tag{7.11}$$

under the Appell replacements of the pairs of Pochammer symbols by a single composite Pochammer symbol.

The four functions known as Appell series (see Srivastava and Karlsson 1985) are

$$F_1(a,\ b,\ b';\ c;\ x,\ y) = \sum_{m,n=0}^{\infty} \frac{(a)_{m+n}(b)_m(b')_n}{(c)_{m+n}} \frac{x^m y^n}{m!n!} \tag{7.12}$$

$$F_2(a, b, b'; c, c'; x, y) = \sum_{m,n=0}^{\infty} \frac{(a)_{m+n}(b)_m(b')_n}{(c)_m(c')_n} \frac{x^m y^n}{m!n!} \tag{7.13}$$

$$F_3(a, a', b, b'; c; x, y) = \sum_{m,n=0}^{\infty} \frac{(a)_m(a')_n(b)_m(b')_n}{(c)_{m+n}} \frac{x^m y^n}{m!n!} \tag{7.14}$$

$$F_4(a, b; c, c'; x, y) = \sum_{m,n=0}^{\infty} \frac{(a)_{m+n}(b)_{m+n}}{(c)_m(c')_n} \frac{x^m y^n}{m!n!}, \tag{7.15}$$

whereas in the case of the Gauss single variable series, the right-hand sides above the double series (7.12)–(7.15) are defined only when the denominator variables c and c' are neither zero nor a negative integer; or when in $c = -\ell$, or $c' = -\ell'$, ℓ, $\ell' = 0, 1, 2, \ldots$ there exists at least one numerator variable, a, a', b, or b', which is a negative integer $-\ell$, such that $k < \min(\ell, \ell')$. Appell's work on the double series was completed by Horn (1931) who found ten more hypergeometric series' in the two variables.

The first attempt at a generalization of the four double series' of Appell—$F_1, \ldots F_4$ in n-variables—is by Lauricella (1893). The status of the subject of multiple Gaussian hypergoemetric series' is systematically presented in Srivastava and Karlsson (1985).

The triple sum series of Jucys and Bandzaitis (1977) for the 9-j coefficient has been identified by Srinivasa Rao and Rajeswari (1989), with a triple hypergeometric series (of unit arguments). Even though there is no resolution to the notational difficulties involved in the study of the multiple hypergeometric series (see Srivastava and Karlsson 1985, p 270); the notation introduced by Srivastava (1967) is resorted to, which unified the various triple hypergeometric functions:

$$F(3)\left[\begin{matrix}(a):: & (b); (b'); (b''); & (c); (c'); (c''); & x, y, z \\ (e):: & (f); (f'); (f''); & (g); (g'); (g''); & \end{matrix}\right]$$
$$= \frac{((a), m+n+p)((b), m+n)((b'), n+p)((b''), p+m)}{((e), m+n+p)((f), m+n)((f'), n+p)((f''), p+m)} \tag{7.16}$$
$$\times \frac{((c), m)((c'), n)((c''), p)}{((g), m)((g'), n)((g'', p))} \frac{x^m y^n z^p}{m!n!p!},$$

where the Pochammer symbol for typographical felicity (see Exton 1976) is written as

$$(\lambda, k) \equiv (\lambda)_k = \lambda(\lambda + 1)\cdots(\lambda + k - 1), \quad k \geqslant 0, \tag{7.17}$$

$$(\lambda, -k) = \frac{(-1)^k}{(1 - \lambda, k)}, \quad k < 0. \tag{7.18}$$

A scrutiny of (7.16) will reveal the role of the colons and the semi-colons as delimiters for the different types of terms in the triple sum series, which depend on either the indices m, n, p, or their linear combinations.

7.3 Definitions of the 9-j coefficient

In atomic and nuclear physics problems, for a two electron—or a two nucleon—problem, with the orbital and spin angular momenta being $\vec{\ell_1}$, $\vec{s_1}$, $\vec{\ell_2}$, $\vec{s_2}$, we come across two coupling schemes called the LS-coupling (or, Russell–Saunders-coupling) and the jj-coupling. In the LS-coupling scheme, the orbital and spin angular momenta are added separately as

$$\vec{\ell_1} + \vec{\ell_2} = \vec{L}, \quad \text{and} \quad \vec{s_1} + \vec{s_2} = \vec{S}. \tag{7.19}$$

In the jj-coupling scheme, which becomes operational in the presence of a spin–orbit interaction, the orbital and spin angular momentum of each electron or nucleon is added separately to give a total angular momentum to each particle as

$$\vec{\ell_1} + \vec{s_1} = \vec{j_1}, \quad \text{and} \quad \vec{\ell_2} + \vec{s_2} = \vec{j_2}. \tag{7.20}$$

The total angular momentum of the two particle system is then the sum of the four angular momenta:

$$\vec{\ell_1} + \vec{s_1} + \vec{\ell_2} + \vec{s_2} = \vec{J}. \tag{7.21}$$

In terms of the LS-coupling (7.19) and the jj-coupling (7.20) schemes, we will have instead of (7.21)

$$\vec{L} + \vec{S} = \vec{J} \tag{7.22}$$

and

$$\vec{j_1} + \vec{j_2} = \vec{J}, \tag{7.23}$$

respectively. We thus have the two coupling schemes for the addition of four angular momenta in the LS-coupling scheme as

$$\vec{\ell_1} + \vec{\ell_2} = \vec{L}, \qquad \vec{s_1} + \vec{s_2} = \vec{S}, \qquad \vec{L} + \vec{S} = \vec{J} \tag{7.24}$$

and in the jj-coupling scheme

$$\vec{\ell_1} + \vec{s_1} = \vec{j_1}, \qquad \vec{\ell_2} + \vec{s_2} = \vec{j_2}, \qquad \vec{j_1} + \vec{j_2} = \vec{J}. \tag{7.25}$$

The complete set of commuting generators required to represent these two coupling schemes are therefore

$$\ell_1^2, \ell_2^2, L^2, s_1^2, s_2^2, S^2, J^2, J_z \tag{7.26}$$

and

$$\vec{\ell}_1^{\,2},\ \vec{s}_1^{\,2},\ \vec{j}_1^{\,2},\ \vec{\ell}_2^{\,2},\ \vec{s}_2^{\,2},\ \vec{j}_2^{\,2},\ \vec{J}^{\,2},\ \vec{J}_z. \tag{7.27}$$

These provide the basis for the orthonormal basis vectors for the *LS* and *jj* coupling schemes, which are denoted as

$$|(\ell_1\ell_2)L(s_1s_2)SJM\rangle \quad \text{and} \quad |(\ell_1s_1)j_1(\ell_2s_2)j_2JM\rangle, \tag{7.28}$$

respectively. These two orthonormal basis vectors are related to each other by an orthogonal transformation:

$$|(\ell_1\ell_2)L(s_1s_2)SJM\rangle = \sum_{j_1,j_2}\begin{pmatrix}\ell_1 & s_1 & j_1\\ \ell_2 & s_2 & j_2\\ L & S & J\end{pmatrix}|(\ell_1s_1)j_1(\ell_2s_2)j_2JM\rangle. \tag{7.29}$$

The orthogonal transformation that relates two uncoupled angular momentum states $|j_1m_1\rangle$ and $|j_2m_2\rangle$ to their coupled state $|JM\rangle$ is

$$|(j_1j_2)JM\rangle = \sum_{m_1,m_2} C(j_1j_2J;\ m_1m_2M)|j_1m_1\rangle|j_2m_2\rangle \tag{7.30}$$

where $C(j_1j_2J;\ m_1m_2M)$ is the Clebsch–Gordan transformation coefficient (also synonymously referred to as the *vector coupling* or *vector addition* coefficients). Using (7.30), it can be shown that the state $|(\ell_1\ell_2)L(s_1s_2)SJM\rangle$ can be decomposed into a sum over the product of three Clebsch–Gordan coefficients multiplied by the four decoupled states $|\ell_1\mu_1\rangle|\ell_2\mu_2\rangle|s_1\nu_1\rangle|s_2\nu_2\rangle$. Then, using the inverse of the transformation (7.30)

$$|j_1m_1\rangle|j_2m_2\rangle = \sum_{J,M} C(j_1j_2J;\ m_1m_2M)|(j_1j_2)JM\rangle, \tag{7.31}$$

we can recouple these states to get the state $|(\ell_1s)j_1(\ell_2s_2)j_2JM\rangle$. The orthonormal states (7.28) are thus related through an orthogonal transformation

$$|(\ell_1\ell_2)L)(s_1s_2)SJM\rangle = \sum_{j_1,j_2}\begin{pmatrix}\ell_1 & s_1 & j_1\\ \ell_2 & s_2 & j_2\\ L & S & J\end{pmatrix}|(\ell_1s_1)j_1(\ell_2s_2)j_2JM\rangle \tag{7.32}$$

where the coefficient in the expansion on the right-hand side of (7.32) is the recoupling coefficient—called the $LS - jj$ transformation coefficient—which, when explicitly written, is the product of six Clebsch–Gordan coefficients summer over three (of the nine) independent projection quantum numbers. This $LS - jj$ transformation coefficient introduced by Wigner (1940) is related to the 9-*j* coefficient as

$$\begin{Bmatrix}\ell_1 & s_1 & j_1\\ \ell_2 & s_2 & j_2\\ L & S & J\end{Bmatrix} = \{[j_1][j_2][J][M]\}^{-1}\begin{pmatrix}\ell_1 & s_1 & j_1\\ \ell_2 & s_2 & j_2\\ L & S & J\end{pmatrix}, \tag{7.33}$$

where $[\lambda] = (2\lambda + 1)^{1/2}$. The unitary property of the recoupling transformation on four angular momenta implies the following orthogonality property for the 9-j coefficient

$$\sum_{j_1, j_2} \begin{Bmatrix} \ell_1 & s_1 & j_1 \\ \ell_2 & s_2 & j_2 \\ L & S & J \end{Bmatrix} \begin{Bmatrix} \ell_1 & s_1 & j_1 \\ \ell_2 & s_2 & j_2 \\ L' & S' & J \end{Bmatrix} [j_1][j_2][L][S] = \delta_{LL'}\delta_{SS'}. \tag{7.34}$$

In terms of the 3-j coefficient, the explicit expression for the 9-j coefficient (Wigner 1940, Edmonds 1957) is

$$\begin{Bmatrix} j_1 & j_2 & j_{12} \\ j_3 & j_4 & j_{34} \\ j_{13} & j_{24} & J \end{Bmatrix} = \sum_{all\ m's} \begin{pmatrix} j_1 & j_2 & j_{12} \\ m_1 & m_2 & m_{12} \end{pmatrix} \begin{pmatrix} j_1 & j_2 & j_{12} \\ m_1 & m_2 & m_{12} \end{pmatrix} \times \begin{pmatrix} j_{12} & j_{34} & J \\ m_{12} & m_{34} & M \end{pmatrix} \begin{pmatrix} j_1 & j_3 & j_{13} \\ m_1 & m_3 & m_{13} \end{pmatrix} \times \begin{pmatrix} j_2 & j_4 & j_{24} \\ m_2 & m_4 & m_{24} \end{pmatrix} \begin{pmatrix} j_{13} & j_{24} & J \\ m_{13} & m_{24} & M \end{pmatrix} \tag{7.35}$$

where the nine angular momenta in the 9-j coefficient satisfy the triangle inequalities satisfied by the six triads

$$(j_1 j_2 j_{12}),\ (j_3 j_4 j_{34}),\ (j_{12} j_{34} J),\ (j_1 j_3 j_{13}),\ (j_2 j_4 j_{24}),\ (j_{13} j_{24} J). \tag{7.36}$$

Wigner (1940) has shown that the 9-j coefficient can also be expressed as a product of three 6-j coefficients summed over a single index, as

$$\begin{Bmatrix} j_1 & j_2 & j_{12} \\ j_3 & j_4 & j_{34} \\ j_{13} & j_{24} & J \end{Bmatrix} = \sum_x (-1)^{2x}(2x+1) \begin{Bmatrix} j_1 & j_3 & j_{13} \\ j_{24} & J & x \end{Bmatrix} \times \begin{Bmatrix} j_2 & j_4 & j_{24} \\ j_3 & x & j_{34} \end{Bmatrix} \begin{Bmatrix} j_{12} & j_{34} & J \\ x & j_1 & j_2 \end{Bmatrix} \tag{7.37}$$

with

$$X_{\min} \leqslant x \leqslant X_{\max} \tag{7.38}$$

$$X_{\min} = \max\{|j_1 - J|, |j_3 - j_{24}|, |j_2 - j_{34}|\} \tag{7.39}$$

$$X_{\max} = \min\{|j_1 + J|, |j_3 + j_{24}|, |j_2 + j_{34}|\}. \tag{7.40}$$

The equation (7.31) also follows from the fundamental theorem of recoupling theory (see Biedenharn and Louck, 1981) according to which every recoupling coefficient 3n-j, $n = 3, 4, \ldots$ is expressible as a sum over a product of the 6-j coefficients.

7.4 Symmetries of the 9-j coefficient

When the 9-j coefficient is written as (7.31), it does not reveal its symmetry properties. Wigner (1940) has indicated that the symmetries of the 9-j coefficient are evident from (7.29), or from the symmetries of the 3-j coefficient. The symmetries of the 9-j coefficient are:

1. the 9-j coefficient is invariant to even column and row permutations. However,
2. for an odd permutation of the columns of the 9-j coefficient results in odd column permutations of three of the six 3-j coefficients in (7.29), which contribute to an overall phase factor of

$$(-1)^{\sigma}, \text{ where } \sigma = j_1 + j_2 + j_3 + j_4 + j_{12} + j_{34} + j_{13} + j_{24} + J, \tag{7.41}$$

3. the 9-j coefficient is invariant under the transposition of its elements, that is

$$\begin{Bmatrix} j_1 & j_2 & j_{12} \\ j_3 & j_4 & j_{34} \\ j_{13} & j_{24} & J \end{Bmatrix} = \begin{Bmatrix} j_1 & j_3 & j_{13} \\ j_2 & j_4 & j_{24} \\ j_{12} & j_{34} & J \end{Bmatrix}. \tag{7.42}$$

Thus, the symmetry group of the 9-j coefficient has 72 elements, is a product of the three columns, and has the three transpositions and transposition, viz $S_3 \times S_3 \times S_2$.

When any one of the nine angular momenta are zero, the 9-j coefficient, from (7.37), can be shown to reduce to a 6-j coefficient. This property can be stated explicitly as

$$\begin{aligned} \begin{Bmatrix} 0 & e & e \\ f & d & b \\ f & e & a \end{Bmatrix} &= \begin{Bmatrix} e & 0 & e \\ c & f & a \\ d & f & b \end{Bmatrix} = \begin{Bmatrix} f & f & 0 \\ d & c & e \\ b & a & e \end{Bmatrix} \\ &= \begin{Bmatrix} f & b & d \\ 0 & e & e \\ f & a & c \end{Bmatrix} = \begin{Bmatrix} a & f & c \\ e & 0 & e \\ b & f & d \end{Bmatrix} = \begin{Bmatrix} b & a & e \\ f & f & 0 \\ d & c & e \end{Bmatrix} \\ &= \begin{Bmatrix} e & d & c \\ e & b & a \\ 0 & f & f \end{Bmatrix} = \begin{Bmatrix} c & e & d \\ a & e & b \\ f & 0 & f \end{Bmatrix} = \begin{Bmatrix} a & b & e \\ c & d & e \\ f & f & 0 \end{Bmatrix} \\ &= \frac{(-1)^{b+c+e+f}}{[e][f]} \begin{Bmatrix} a & b & e \\ d & c & f \end{Bmatrix}. \end{aligned} \tag{7.43}$$

7.5 The Jucys–Bandzaitis triple sum series

The conventional single sum expression for the 9-j coefficient (7.31), which can be derived from the fundamental theorem of recoupling theory, is in fact a sum over four variables, since each one of the three 6-j coefficients, which occur in (7.31), have been defined as sets of generalized hypergeometric ${}_4F_3(1)$ functions of the unit arguments (6.31) and (6.42).

The simplest known algebraic form for the 9-j coefficient is the triple sum series, from Jucys and Bandzaitis (1977):

$$\begin{Bmatrix} a & b & c \\ d & e & g \\ g & h & i \end{Bmatrix} = (-1)^{z5} \frac{(beh)(dag)(igh)}{((bac)(def)(icf)} \times \sum_{x,y,z} \frac{(-1)^{x+y+z}}{x!y!z!} \frac{(x1-x)!(x2+x)!(x3+x)!}{(x4-x)!(x5-x)!} \times \frac{(y1+y)!(y2+y)!}{(y3+y)!(y4-y)!(y5-y)!} \frac{(z1-z)!(z2+z)!}{(z3-z)!(z4-z)!(z5-z)!} \times \frac{(p1-y-z)!}{(p2+x+y)!(p3+x+z)!} \tag{7.44}$$

where

$$0 \leqslant x \leqslant \min(-d+e+f,\, c+f-i) = XF, \tag{7.45}$$

$$0 \leqslant y \leqslant \min(g-h+i,\, b+e-h) = YF, \tag{7.46}$$

$$0 \leqslant z \leqslant \min(a-b+c,\, a+d-g) = ZF, \tag{7.47}$$

$$x1 = 2f, \quad y1 = -b+e+h, \quad z1 = 2a, \tag{7.48}$$

$$x2 = d+e-f, \quad y2 = g+h-i, \quad z2 = -a+b+c, \tag{7.49}$$

$$x3 = c-f+i, \quad y3 = 2h+1, \quad z3 = a+d+g+1, \tag{7.50}$$

$$x4 = -d+e+f, \quad y4 = b+e-h \quad z4 = a+d-g, \tag{7.51}$$

$$x5 = c+f-i, \quad y5 = g-h+i, \quad z5 = a-b+c, \tag{7.52}$$

$$p1 = a + d - h + i, \quad p2 = -b + d - f + h, \quad p3 = -a + b - f + i \tag{7.53}$$

$$(abc) = \left[\frac{(a - b + c)!(a + b - c)!(a + b + c + 1)!}{(-a + b + c)!}\right]^{\frac{1}{2}}, \tag{7.54}$$

and, as before, $\Gamma(x, y, \ldots) = \Gamma(x)\Gamma(y)\ldots$.

The single sum over a product of three 6*j*-coefficients for the 9-*j* coefficient (7.37), and the fact that the 6-*j* coefficient can be expressed as ${}_4F_3(1)$, studied in chapter 6, suggests that, to understand the triple sum series for the 9-*j* coefficient, the starting point is the product of the following ${}_4F_3(1)$ functions

$$\begin{aligned}
&{}_4F_3\begin{pmatrix} 1 + x2, 1 + x3, -x4, -x5; 1 \\ -x1, 1 + p2, 1 + p3 \end{pmatrix} \\
&\times {}_4F_3\begin{pmatrix} 1 + y1, 1 + y2, -y4, -y5; 1 \\ 1 + y3, -p1, 1 + p2 \end{pmatrix} \\
&\times {}_4F_3\begin{pmatrix} 1 + z2, -z3, -z4, -z5; 1 \\ -z1, -p1, 1 + p2 \end{pmatrix} = \sum_{x,y,z} \frac{1}{x!y!z!} \\
&\times \frac{(1 + x2, x)(1 + x3, x)(-x4, x)(-x5, x)}{(-p1, x)(1 + p2, x)(1 + p3, x)} \\
&\times \frac{(1 + y1, y)(1 + y2, y)(-y4, y)(-y5, y)}{(1 + y3, y)(-p1, y)(1 + p2, y)} \\
&\times \frac{(1 + z2, z)(-z3, z)(-z4, z)(-z5, z)}{(-z1, z)(-p1, z)(1 + p3, z)}.
\end{aligned} \tag{7.54}$$

Akin to what Appell (1925) did to create double sum series out of a product of two single sum series, in (7.54) we make the following ad hoc replacements of three pairs of Pochammer symbols

$$(1 + p2, x)(1 + p2, y) \quad \text{by} \quad (1 + p2, x + y) \tag{7.55}$$

$$(1 + p3, x)(1 + p3, z) \quad \text{by} \quad (1 + p3, x + z) \tag{7.56}$$

$$(-p1, y)(-p1, z) \quad \text{by} \quad (-p1, y + z) \tag{7.57}$$

to identify (7.54) with the triple hypergeometric series

$$\begin{aligned}
&F^{(3)}\left(\begin{matrix} -:: & -; & -; & -: & 1 + x2, 1 + x3, -x4, -x5; \\ -:: & 1 + p2; & -p1; & 1 + p3: & -x1 \end{matrix}\right. \\
&\left.\begin{matrix} 1 + y1, 1 + y2, -y4, -y5; & 1 + z2, -z3, -z4, -z5 & 1, 1, 1 \\ 1 + y3; & -z1; & \end{matrix}\right).
\end{aligned} \tag{7.58}$$

Equation (7.58) is clearly a special case of the extremely general function defined in three variables by Srivastava (1967), which is an elegant unification (Exton 1976) of the triple hypergeometric functions of Lauricella (1893), Saran (1954), and Srivastava (1964) functions, which we call the Lauricella–Saran–Srivastava triple hypergeometric functions.

The numerical values of the 18 parameters given in (7.48)–(7.53) are different for different symmetries of the 9-j coefficient, since they are dependent on the positions of $a, b, \ldots, i$ in the 3×3 array of real numbers. They are also dependent since there are nine relations between them:

$$\begin{gathered} x2 + x4 = y1 + y4, \quad x3 + x5 = z2 + z5, \\ y2 + y5 + 1 = z3 - z4, \\ x2 = y4 + p2, \quad x3 = p3 + z5, \quad y1 = x4 + p2, \\ p1 = y5 + z4, \quad p2 = x2 - y4, \quad p3 = z2 - x5. \end{gathered} \tag{7.59}$$

Bandzaitis, Karosienne, and Jucys (1964) derived formulae for the stretched 9-j coefficients. The observation that a given 9-j coefficient may have one or more terms has led Srinivasa Rao and Van der Jeugt (1994) to search for the stretched 9-j coefficient formulae, and the triple sum series for the summation theorems. They showed that the Vandermonde theorem for the ${}_2F_1(1)$, the Pfaff–Saalschütz summation theorem of ${}_3F_2(1)$, and the Karlsson (1971)–Minton summation theorem for the ${}_4F_3(1)$ occur naturally in their study. Besides these well known theorems, the study opened up the scope for finding new summation theorems.

The lack of symmetry is best illustrated through a numerical example:

$$\begin{Bmatrix} 30 & 20 & 10 \\ 30 & 10 & 20 \\ 60 & 30 & 30 \end{Bmatrix} = 0.00026845, \tag{7.60}$$

which has $XF = 0$, $YF = 0$, $ZF = 0$, so that it is a single term. But, its symmetry (a cyclic column permutation followed by an odd row permuatation of the 9-j treated as a 3×3 array)

$$\begin{Bmatrix} 30 & 20 & 10 \\ 30 & 10 & 20 \\ 60 & 30 & 30 \end{Bmatrix} = 0.00026845 \tag{7.61}$$

has $XF = 60$, $YF = 20$, $ZF = 40$, and so has an actual number of 33 761 terms—a number determined by taking into account the constraints on the ranges of the summation indices x, y, z, by the parameters $p1$, $p2$, $p3$, viz

$$y + z \leqslant p1, \quad \text{and if} \quad p2, p3 < 0, \quad \text{then} \quad x + y \geqslant |p2|, \quad x + z \geqslant |p3|. \tag{7.62}$$

If we were to use the expression from Wigner (7.37) for the 9-j coefficient, where it is written as a single sum over a product of three 6-j coefficients; for the 6-j coefficient itself to be given by the single sum expression from Racah (see equation (6.9)), the 9-j coefficient would require 40 terms; that is 120 values of the 6-j coefficient (since each term is a product of three 6-*j*s). But the 9-*j* written as a triple sum series (7.44) would require one term, or only three values of the 6-*j* coefficient. From this, it can be argued that the triple sum series (7.44) requires less time for its numerical computation than the conventional Wigner single sum over a product of 6-*j* coefficients, especially for large values of the angular momenta.

7.6 Stretched 9-*j* coefficients

A 9*j*-coefficient is called *stretched* if in any one of the six triads (7.36) (*abc*, say) is one or more of the angular momenta belonging to the triad $|a - b| \leqslant c \leqslant (a + b)$. That is

$$\begin{aligned}&(a\ b\ a+b),\ (a\ a+b\ b),\ \text{or}\ (a+b\ a\ b),\ \text{and}\\&(a\ b\ a-b),\ (a\ a-b\ b),\ \text{or}\ (a-b\ a\ b)\end{aligned} \tag{7.63}$$

corresponds to the limits of the triangle inequality. Sharp (1967) showed that there are, in all, five distinctly different doubly stretched cases and two triply stretched 9-*j* coefficients, while any singly stretched 9-*j* coefficient can be brought to one standard form through the symmetries of the coefficient. The five distinct types of doubly stretched 9-*j* coefficients defined by Sharp (1967) are:

$$\text{(I)}\quad \begin{Bmatrix} a & b & c \\ d & e & f \\ a+d & b+e & i \end{Bmatrix} \text{corresponding to } y4 = 0,\ z4 = 0 \tag{7.64}$$

$$\text{(II)}\quad \begin{Bmatrix} a & b & c \\ d & d+f & f \\ a+d & h & i \end{Bmatrix} \text{corresponding to } e = d+f,\ z4 = 0 \tag{7.65}$$

$$\text{(III)}\quad \begin{Bmatrix} a & b & c \\ d & e & f \\ a+d & a+d+i & i \end{Bmatrix} \text{corresponding to } y5 = 0,\ z4 = 0 \tag{7.66}$$

$$\text{(IV)}\quad \begin{Bmatrix} a & b & c \\ d & b+h & f \\ a+d & h & i \end{Bmatrix} \text{corresponding to } e = b+h,\ z4 = 0 \tag{7.67}$$

$$\text{(V)}\quad \begin{Bmatrix} a & b & c \\ d & e & f \\ a+d=h+i & h & i \end{Bmatrix} \text{corresponding to } g = h+i,\ z4 = 0. \tag{7.68}$$

Any doubly stretched 9-j coefficient can be classified into one of these five types using the well known symmetries—column, row permutations, and transposition about the leading diagonal—of the coefficient. Explicit formulae for these five types have been derived by Sharp (1967) from the expression for the singly stretched 9-j coefficient,

$$\begin{Bmatrix} a & b & c \\ d & e & f \\ a+d & b+e & i \end{Bmatrix} \text{corresponding to } z4 = 0 \text{ or } g = a + d. \tag{7.69}$$

The formulae from Sharp are single sum series, except for type (III), which is a single term. From the fact that $z4 = 0$ for all five types and, in addition, $y4 = 0$ for type (I) and $y5 = 0$ for type (III), it follows that the triple-sum series reduces to a double-sums series for types (II), (IV), and (V), and to a single-sum for (I) and (III). However, this is not the complete picture.

We will analyze the type (III) doubly stretched 9-j coefficient through its triple-sum series representation and the consequences of some of the symmetries on it.

Case (i). When we set $y5 = 0$ and $z4 = 0$ in the triple-sum series, we get, for the type (III) doubly stretched 9-j coefficient, the expression

$$\begin{aligned}\begin{Bmatrix} a & b & c \\ d & e & f \\ a+d & a+d+i & i \end{Bmatrix} &= (-1)^{z5}\frac{(d,\, a,\, a+d)\,(b,\, e,\, a+d+i)}{(b,\, a,\, c)\,(d,\, e,\, f)} \\ &\times \frac{(i,\, a+d,\, a+d+i)}{(i,\, c,\, f)}\frac{y1!\; y2!}{y3!\; y4!}\frac{z1!z2!}{z3!\; z5!} \\ &\times \sum_x \frac{(-1)^x}{x!}\frac{(x1-x)!\,(x2+x)!\,(x3+x)!}{(x4-x)!\,(x5-x)!\,(p2+x)!\,(p3+x)!}.\end{aligned} \tag{7.70}$$

Interestingly, the triple-sum series for a symmetry of these 9-j coefficients (see below) appears to be a double sum, since it has $z5 = 0$ and hence, $z = 0$. However, if the additional constraints on the summation indices (7.62) are taken into account; then, since

$$\begin{aligned} &x \leqslant x5(=c-f+i), \text{ and } y \leqslant y5(=d-e+f), \\ &x + y \geqslant -p2(=c+d-e+i), \end{aligned} \tag{7.71}$$

the apparent double-sum reduces to a single term and we have the result

$$\begin{aligned}\begin{Bmatrix} a+d & a+d+i & i \\ a & b & c \\ d & e & f \end{Bmatrix} &= (-1)^{d-e+f}\frac{(a+d+i,\, b,\, e)}{(a,\, b,\, c)\,(d,\, e,\, f)\,(i,\, c,\, f)} \\ &\times \left[\frac{(2a)!\,(2d)!\,(2i)!}{(2a+2d+1)\,(2a+2d+2i+1)!}\right]^{\frac{1}{2}}.\end{aligned} \tag{7.72}$$

This result has been derived by Sharp from the formula for the singly stretched 9-j coefficient in terms of a double sum given by Sharp and von Baeyer (1966).

When the single-sum series, part of (7.70), is rearranged into a generalized hypergeometric function, we have

$$S_x = \frac{x1!\,x2!\,x3!}{x4!\,x5!\,p2!\,p3!}\,{}_4F_3\left(\begin{matrix}1+x2,\ 1+x3,\ -x4,\ -x5\\ -x1,\ 1+p2,\ 1+p3\end{matrix}\right), \tag{7.73}$$

where ${}_4F_3(1)$ is not Saalschützian and therefore cannot be a Racah or 6-j coefficient (Srinivasa Rao and Rajeswari 1993). However, it is a zero-balanced ${}_{p+1}F_p(1)$ with equal numerator and denominator variable sums:

$$\begin{aligned}&1+x2+1+x3-x4-x5=-x1+1+p2+1+p3\\ \text{or,}\qquad &x1+x2+x3=x4+x5+p2+p3=c+d+e+i,\end{aligned} \tag{7.74}$$

using the values of $x1$, $x2$, $x3$, $x4$, $x5$, $p2$, $p3$ given in (7.48)–(7.53).

For a comparison of these results for the two symmetries of the doubly stretched 9-j coefficient, after simplification—for the summation part—the result is

$$S_x = (-1)^{x1-x4-x5}, \tag{7.75}$$

which, when expressed in terms of ${}_4F_3(1)$, replaces the factorials by gamma functions and uses the notation for the product of gamma functions $\Gamma(a, b, \ldots) = \Gamma(a)\Gamma(b)\ldots$, which yields

$$\begin{aligned}&{}_4F_3\left(\begin{matrix}1+x2,\ 1+x3,\ -x4,\ -x5\\ -x1,\ 1+p2,\ 1+p3\end{matrix}\right) = (-1)^{x1-x4-x5}\\ &\times\frac{\Gamma(1+x4,\ 1+x5,\ 1+p2,\ 1+p3)}{\Gamma(1+x1,\ 1+x2,\ 1+x3)}.\end{aligned} \tag{7.76}$$

We now make the identifications

$$-x1 = b1,\ -x4 = b1 + m1, \qquad (m1 = x1 - x4), \tag{7.77}$$

$$1 + p2 = b2, \qquad 1 + x2 = b2 + m2, \qquad (m2 = x2 - p2), \tag{7.78}$$

$$1 + p3 = b3, \qquad 1 + x3 = b3 + m3, \qquad (m3 = x3 - p3), \tag{7.79}$$

and we rewrite the result as

$$\begin{aligned}&{}_4F_3\left(\begin{matrix}-(m1+m2+m3),\ b1+m1,\ b2+m2,\ b3+m3\\ b1,\qquad b2,\qquad b3\end{matrix}\right) = (-1)^{m1+m2+m3}\\ &\times\frac{(m1+m2+m3)!}{(b1)_{m1}\,(b2)_{m2}\,(b3)_{m3}},\end{aligned} \tag{7.80}$$

where $(b)_m = \Gamma(b + m)/\Gamma(b)$ is the Pochammer symbol.

Equation (7.80) is the Karlsson–Minton summation theorem for the terminating zero-balanced ${}_{p+1}F_p(1)$ series corresponding to $p = 3$ (cf (1.9.3) of Gasper and Rahman 1991).

Case(ii). For another symmetry of the 9-j coefficient considered in case (i), the triple sum series yields the result

$$\begin{Bmatrix} c & b & a \\ f & e & d \\ i & a+d+i & a+d \end{Bmatrix} = \frac{(f, c, i)\,(b, e, a+d+i)\,(a+d, i, a+d+i)}{(f, e, d)\,(b, c, a)\,(a+d, a, d)} \\ \times \frac{(x1)!(x2)!(x3)!}{(x4)!} \frac{(y1)!(y2)!}{(y3)!(y4)!} \frac{(z1)!}{(z3)!(z5)!(p2)!} {}_2F_1(-z3, -z4; -z1; 1), \tag{7.81}$$

where

$$x1 = 2d, \qquad y1 = a - b + d + e + i, \qquad z1 = 2c, \tag{7.82}$$

$$x2 = -d + e + f, \qquad y2 = 2i, \qquad z2 = a + b - c, \tag{7.83}$$

$$x3 = 2a, \qquad y3 = 2(a + d + i) + 1, \qquad z3 = c + f + i + 1, \tag{7.84}$$

$$x4 = d + e - f, \qquad y4 = -a + b - d + e - i, \qquad z4 = c + f - i, \tag{7.85}$$

$$x5 = 0, \qquad y5 = 0, \qquad z5 = a - b + c, \tag{7.86}$$

$$p1 = c + f - i, \qquad p2 = a - b + f + i, \qquad p3 = a + b - c. \tag{7.87}$$

Comparison with the column-permuted result in (7.72) yields immediately after simplification. Using the symmetry property of the 9-j coefficient, the result is

$$\begin{aligned} {}_2F_1(-z3, -z5; -z1; 1) &= (-1)^{-a+b-c} \frac{(-a+b+c)!\,(a-b+f+i)!}{(-c+f-i)!(2c)!} \\ &= \frac{(-z1+z3)_{z5}}{(-z1)_{z5}}, \end{aligned} \tag{7.88}$$

which is clearly a manifestation of the Vandermonde summation theorem,

$${}_2F_1(a, -n; c; 1) = \frac{(c-a)_n}{(c)_n}. \tag{7.89}$$

Case (iii). We consider another symmetry of the 9-j coefficient for which the triple-sum series again reduces to a single-sum series:

$$\begin{Bmatrix} b & c & a \\ a+d+i & i & a+d \\ e & f & d \end{Bmatrix} = (-1)^{x5+z5} \frac{(a+d+i,\, b,\, e)}{(b,\, a,\, c)\,(a+d+i,\, i,\, a+d)}$$
$$\times \frac{(c,\, i,\, f)(d,\, e,\, f)}{(c,\, b,\, a)(d,\, a,\, a+d)} \frac{x1!\, x2!\, x3!}{x4!\, x5!} \frac{(z1-z5)!\,(z2+z5)!}{z5!\,(z3-z5)!\,(z4-z5)!\,(p3+z5)!} \qquad (7.90)$$
$$\times \sum_y \frac{(-1)^y}{y!} \frac{(y2+y)!\,(p1-z5-y)!}{(y3+y)!\,(y4-y)!\,(y5-y)!}.$$

Comparison with a symmetry (cyclic-column permutation with an odd-row permutation) of (7.72), yields, on simplification, the result

$$_3F_2(1+y2,\, -y4,\, -y5;\, -1+y3,\, -p1+z5;\, 1)$$
$$= \frac{\Gamma(-y2+y3+y4,\, 1+y3+y4+y5,\, 1+y3,\, -y2+y3+y4)}{\Gamma(-y2+y3,\, 1+y3+y4,\, 1+y3+y5,\, -y2+y3+y4+y5)}, \qquad (7.91)$$

using the Saalschütz property satisfied by the variables of the $_3F_2(1)$, that is

$$1+y2-y4-y5 = y3+z5-p1. \qquad (7.92)$$

This formula (7.92) can be shown to be the Pfaff–Saalschütz summation theorem (see Gasper and Rahman 1991):

$$_3F_2(a,\, b,\, -n;\, c,\, 1+a+b-c-n;\, 1) = \frac{(c-a)_n\,(c-b)_n}{(c)_n\,(c-a-b)_n}. \qquad (7.93)$$

Case (iv). Next, we consider a symmetry of the type (III) 9-j coefficient for which the triple sum series reduces to a double-sum series

$$\begin{Bmatrix} f & i & c \\ e & a+d+i & b \\ d & a+d & a \end{Bmatrix} = \frac{(e,\, d,\, f)\,(i,\, a+d+i,\, a+d)\,(a,\, d,\, a+d)}{(e,\, a+d+i,\, b)\,(i,\, f,\, c)\,(a,\, c,\, b)}$$
$$\times \frac{y1!\, y2!}{y3!\, y4!\, y5!} \sum_{x,z} \frac{(-1)^{x+z}}{x!\, z!} \frac{(x1-x)!\,(x2+x)!\,(x3+x)!}{(x4-x)!\,(x5-x)!} \qquad (7.94)$$
$$\times \frac{(z5-z)!\,(z2+z)!}{(z3-z)!\,(z5-z)!\,(p3+z)!\,(p2+x+z)!}.$$

Comparison of this result with a symmetry (an odd-column permutation, a cyclic-row permutation, and a reflection on the diagonal) of the single-sum expression (7.92), after simplification, yields the result for the double series

$$\sum_{x,z} \frac{(-1)^{x+z}}{x!z!} \frac{(x1-x)!\,(x2+x)!\,(x3+x)!}{(x4-x)!\,(x5-x)!(p2+z)!} \frac{(z1-z)!\,(z2+z)!}{(z3-z)!\,(z5-z)!\,(p3+x+z)!}$$
$$= (-1)^{z2+z5} \frac{(x1+x2+1)!\,x2!\,(z1+z2-z5)!}{(x2+x4+1)!\,(x2-p3+1)!\,(x2-p3-z)!} \tag{7.95}$$

where

$$x1 = 2b, \qquad z1 = 2f, \tag{7.96}$$

$$x2 = a - b + d + e + i, \qquad z2 = c - f + i, \tag{7.97}$$

$$x3 = a - b + c, \qquad z3 = d + e + f + 1, \tag{7.98}$$

$$x4 = a + b + d - e + i, \qquad z2 = c + f - i, \tag{7.99}$$

$$x5 = -a + b + c, \qquad z2 = a - b - f + i, \tag{7.100}$$

$$p2 = a - b + d + e - i. \tag{7.101}$$

This sum depends on seven independent variables, e.g. $x1$, $x2$, $x4$, $z1$, $z2$, $z5$ and $p3$; the remaining variables are then given by

$$x3 = z5 + p3,\ x5 = z2 - p3,\ z5 = x2 - p3 + 1,\ p2 = x2 - z1 - z2 + z5. \tag{7.102}$$

Furthermore, there are inequality conditions governed by the requirement that the factorials should be non-negative. The factor $(p3 + x + z)!$ in this sum makes it a genuine double-sum series. Numerically, the validity of this summation theorem has been checked with MACSYMA. In addition, the summation in x (or z separately) cannot be performed with the help of other summation theorems (like the Minton 1970 or Karlson–Minton summation theorems, Miller and Srivastava 2010 for ${}_4F_3(1)$). If we consider the symmetry of the 9-j coefficient:

$$\begin{Bmatrix} b & c & a \\ d & e & f \\ a+d+i & i & a+d \end{Bmatrix} \tag{7.103}$$

for which the triple-sum series also reduces to a double sum. It can be summed using first the Minton theorem for ${}_4F_3(1)$—though a numerator parameter contains the other summation index, the characteristic of the Minton theorem is the presence of a numerator and a denominator parameter, which differ by 1—and the resulting series can be summed using the $_3F_2(1)$ summation theorem of Vandermonde.

Case (v). Here we consider a symmetry of the 9-j coefficient, which does not reduce the triple-sum series

$$\begin{Bmatrix} d & e & f \\ a+d & a+d+i & i \\ a & b & c \end{Bmatrix} = (-1)^{x5} \frac{(a+d,d,a)\,(e,a+d+i,b)\,(c,a,b)}{(a+d,a+d+i,i)\,(e,d,f)\,(c,f,i)}$$
$$\times \frac{(-1)^{x+y+z}}{x!\,y!\,z!} \frac{(x2+x)!\,(x3+x)!}{(x5-x)!} \frac{(y1+y)!(y2+y)!}{(y3+y)!\,(y4-y)!)y5-y)!} \qquad (7.104)$$
$$\times \sum_{x,z} \frac{(-1)^{x+z}}{x!\,z!} \frac{(x1-x)!\,(x2+x)!\,(x3+x)!}{(x4-x)!\,(x5-x)!}$$
$$\times \frac{z5-z)!\,(z2+z)!}{(z3-z)!\,(z5-z)!\,(p3+z)!\,(p2+x+z)!},$$

where we choose $x3$, $x5$, yi, $y4$, $z5$, $p2$ to be independent variables. The dependent variables are related to these through

$$x2 = y4 + p2, \qquad y2 = x5 - z5 + p2, \qquad (7.105)$$

$$y5 = x3 + y1 - y3 - z5 + 1, \qquad z2 = x3 + x5 - z5, \qquad (7.106)$$

$$z3 = y4 + p2 + 1, \qquad p1 = -x5 + y4 + z5, \qquad (7.107)$$

$$p3 = x3 - z5. \qquad (7.108)$$

The numerical validity of the summation theorems has also been verified using the symbolic software package Macsyma. This triple-sum cannot be summed with the help of the Minton theorem.

Thus, we have shown that the comparison of a stretched 9-j coefficient formula with the Jucys–Bandzaitis triple sum series, in conjunction with the symmetries of the 9-j coefficient, reveals the Vandermonde, Pfaff–Saalschütz, and Karlsson–Minton summation theorems for the $_2F_1(1)$, $_3\mathrm{F} - 2(1)$, $_4F_3(1)$ series, as well as new summation theorems for double and triple-sum series'. This is a direct consequence of the highly asymmetric nature of the triple-sum series. A complete classification of the summation theorems from a study of all 72 symmetries of the 9-j coefficient on the triple-sum series and their relations to the other stretched formulae remains to be done.

7.7 A general transformation formula

A general hypergeometric series in two variables is called a Kampé de Fériet series (1923). For such a series, some isolated summation formulas have been established.

Van der Jeugt and Pitre (1996) proposed a new approach for deriving a number of transformation and summation formulas, and this has been extended by Van der Jeugt, Pitre, and Srinivasa Rao (1997). General Kampé de Fériet hypergoemetric series' (1923) in two variables are defined as follows:

$$F^{A:B;B'}_{C:D;D'}\left[\begin{matrix}(a):(b);(b');\\(c):(d);(d');\end{matrix}x,y\right]=\sum_{k,\ell=0}^{\infty}\frac{\prod_{j=1}^{A}(a_j)_{k+\ell}\prod_{j=1}^{B}(b_j)_k\prod_{j=1}^{B'}(b'_j)_\ell}{\prod_{j=1}^{C}(c_j)_{k+\ell}\prod_{j=1}^{D}(b_j)_k\prod_{j=1}^{D'}(d'_j)_\ell}\frac{x^k}{k!}\frac{y^\ell}{\ell!}. \quad (7.109)$$

Here, $(a)=(a_1, a_2, \ldots a_A)$, and $(\alpha)_n$ are the class of Pochammer symbols. It is understood that no zeros appear in the denominator. We shall be concerned in this section with the case $B'=B$ and $D'=D$.

Convergence criteria for the above series were studied by Srivastava and Daoust (1972), and by Hai, Marichev, and Srivastava (1992). For us, the case: $x=y=1, A=0$, $C=a$ is of particular interest. In this case, the series converges absolutely provided

$$\mathcal{R}\left(c+\sum_{j=1}^{D}d_j-\sum_{j=1}^{B}b_j\right)>0 \text{ and } \mathcal{R}\left(c+\sum_{j=1}^{D'}d'_j-\sum_{j=1}^{B'}b'_j\right)>0. \quad (7.110)$$

For a double hypergeometric series of the unit argument, there are comparatively fewer transformation and summation formulas available in mathematics literature. Jain (1966) obtained a summation formula for a particular $F^{0:3;3}_{1:1;1}(1,1)$ series, and Carlitz (1967) for a certain $F^{1:2;2}_{0:2;2}(1,1)$ series; these were also studied by Srivastava (1984). Studies of the 9-j coefficient has led us to more transformation and summation formulas for the $F^{0:3;4}_{1:1;2}(1,1)$ series, and the $F^{1:2;2}_{0:2;2}(1,1)$. These results are presented in this section. Of particular interest is the list of summation formulas for the $F^{1:2;2}_{0:2;2}(1,1)$ series (including the one obtained by Carlitz 1967). To start with a transformation formula, relating a $F^{0:3;p+3}_{1:1;p+1}(1,1)$ series to $F^{1:2;p+2}_{0:2;p}(1,1)$, is derived.

Proposition. Let $f_{p+3}=d-a$, $g_{p+1}=d+e-a-b-c$, and

$$\mathcal{R}\left(a+\sum_{j=1}^{p}g_j-\sum_{j=1}^{p+2}f_j\right)>0, \quad \mathcal{R}(f_{p+3})>0, \quad \mathcal{R}(g_{p+1})>0. \quad (7.111)$$

Then

$$\begin{aligned}F^{0:3;p+3}_{1:1;p+1}&\left[\begin{matrix}-:a,\ b,\ c;f_1,f_2,\ldots f_p+3;\\ d:\quad e;g_1,g_2,\ldots g_p+1;\end{matrix}x,y\right]=\Gamma\left[\begin{matrix}d,\ d+e-a-b-c\\ d-a,\ d+e-b-c\end{matrix}\right]\\ &\times F^{0:3;p+2}_{1:1;p}\left[\begin{matrix}-&:\ a,e-b,e-c\ ;\ f_1,f_2,\ldots f_p+2\ ;\\ d+e-b-c&:\quad e\quad ;\quad g,g_2,\ldots g_p\quad ;\end{matrix}1,1\right],\end{aligned} \quad (7.112)$$

where

$$\Gamma\begin{bmatrix} a_1, a_2, \dots \\ b_1, b_2, \dots \end{bmatrix} = \frac{\Gamma(a_1)\Gamma(a_2)\cdots}{\Gamma(b_1)\Gamma(b_2)\cdots}. \tag{7.113}$$

Proof. From the beta function integral representation:

$$B(x, y) = \Gamma\begin{bmatrix} x + y \\ x, y \end{bmatrix} \int_0^1 t^{x-1}(1 - t)^{y-1}dt, \ \mathcal{R}(x) > 0, \ \mathcal{R}(y) > 0, \tag{7.114}$$

we deduce

$$\frac{(x)_k\,(y)_\ell}{(x + y)_{k+\ell}} = \Gamma\begin{bmatrix} x + y \\ x, y \end{bmatrix} \int_0^1 t^{x-1}(1 - t)^{y-1}dt. \tag{7.115}$$

Assume that $\mathcal{R} > 0$, and apply the above formula to $(a)_k(fp + 1)_\ell/(d)_{k+\ell}$ with $f_{p+3} = d - a$. Using the Pochammer symbol convention $(b, c)_k = (b)_k(c)_k$, the left-hand side of (7.112) becomes

$$\Gamma\begin{bmatrix} d \\ a, d - a \end{bmatrix} \int_0^1 \sum_{k,\ell} \frac{(b, c)_k\,(f_1, \dots, f_{p+2})_\ell}{(e)_k k!(g_1, \dots g_{p+1})_\ell \ell!} t^{a+k-1}(1 - t)^{d-a+\ell-1}dt. \tag{7.116}$$

In (7.116), using Euler's identity (see Lucy Slater 1966, equation (1.3.15)):

$$\sum_k \frac{(b, c)_k t^k}{(e)_k k!} = {}_2F_1(b, c; e; t) = (-1)^{e-b-c}{}_2F_1(e - b, e - c; e; t), \tag{7.117}$$

we get, for (7.116):

$$\Gamma\begin{bmatrix} d \\ a, d - a \end{bmatrix} \int_0^1 \sum_{k,\ell} \frac{(e - b, e - c)_k(f_1, \dots f_{p+2})_\ell}{(e)_k k!(g_1, \dots, g_{p+1})_\ell \ell!} t^{a+k-1}(1 - t)^{d+e-a-b-c+\ell-1}dt, \tag{7.118}$$

$$\Gamma\begin{bmatrix} d, d + e - a - b - c \\ d - a, -d + e - b - c \end{bmatrix} \sum_{k,\ell=0}^{\infty} \frac{(a, e - b, e - c)_k(f_1, \dots, f_{p+2}, d + e - a - b - c)_\ell}{(e)_k(g_1, \dots, g_{p+1})_\ell(d + e - b - c)_{k+\ell}k!\,\ell!}, \tag{7.119}$$

leading to the right-hand side (7.112), since $g_{p+1} = d + e - a - b - c$. The absolute convergence conditions are necessary for the interchange of summation and integration; the extra condition $\mathcal{R} > 0$, used to apply the beta function integral, disappears by analytic continuation. QED

7.8 Transformation formulas for $F^{0:3;4}_{1:1;2}$

Pitre and Van der Jeugt (1996) derived three transformation formulas for the double series of the type $F^{0:3;3}_{1:1;1}(1, 1)$.

- For $\mathcal{R}(d+e-a-b-c)>0$, $\mathcal{R}(d-a>0)$ and $\mathcal{R}(d+e-b-c-b'-c')>0$,

$$F^{0:3;p+3}_{1:1;p+1}\left[\begin{matrix} -:a,\ b,\ c;f_1,f_2,\ldots f_p+3; \\ d:\quad e\quad ;g_1,g_2,\ldots g_p+1; \end{matrix} x,y\right]$$
$$=\Gamma\left[\begin{matrix} d,\ d+e-a-b-c,\ d+e-b-c-b'-c' \\ d+e-b-c-b',\ d+e-b-c-c' \end{matrix}\right] \tag{7.120}$$
$$\times\ {}_4F_3\left[\begin{matrix} a,\ e-b,\ e-c,\ d+e-b-c-b'-c' \\ e,\ d+e-b-c-b',\ d+e-b-c-c' \end{matrix}\right],$$

- For n, a non-negative integer, and $\mathcal{R}(e'-a'-c'-n)>0$,

$$F^{0:3;p+3}_{1:1;p+1}\left[\begin{matrix} - & : -n,\ a,\ b & ;\ 1+a+b-e,\ a',\ b' & ; \\ 1+a+b-e-n & :\quad e & ;\quad e' & ; \end{matrix} 1,1\right]$$
$$=\frac{(e-a,\ e-b)_n}{(e,\ e-a-b)_n}\Gamma\left[\begin{matrix} e,\ e'-a'-b' \\ e'-a',\ e'-b' \end{matrix}\right] \tag{7.121}$$
$$\times\ {}_4F_3\left[\begin{matrix} -n,\ a',\ b',\ 1-e-n & ; \\ 1+a-e-n,\ 1+b-e-n,\ 1+a'+b'-e' & ; \end{matrix} 1\right].$$

- For n, a non-negative integer, and $\mathcal{R}(e-a-c-n)>0$,

$$F^{0:3;p+3}_{1:1;p+1}\left[\begin{matrix} - & : a,\ b,\ d+n & ;\ d-a,\ b',\ -n & ; \\ d & :\quad e & ;\quad e' & ; \end{matrix} 1,1\right]$$
$$=\frac{(a)_n}{(b)_n}\Gamma\left[\begin{matrix} e,\ e-a-b-n \\ e-b,\ e-a-n \end{matrix}\right] \tag{7.122}$$
$$\times\ {}_4F_3\left[\begin{matrix} -n,\ d-a,\ e'-b',\ e-a-b-n & ; \\ e',\ 1-a-n,\ e-a-n & ; \end{matrix} 1\right].$$

Next we prove three transformation formulas for the $F^{0:3;4}_{1:1;2}$ series. For each of these, the conditions of proposition (7.111) are satisfied. They reduce to a $F^{0:3;3}_{1:1;2}(1, 1)$ series. Herein, the parameters are such that (7.120), (7.121), or (7.122) are applicable. Thus, we obtain the following results:

Corollory. Let $\Re(f) > 0$, $\Re(f - c) > 0$ and $\Re(e + f - a - b - c) > 0$. Then

$$F^{0:3;4}_{1:1;2}\left[\begin{matrix} - : a, b, c ; & e - a, e - b, c', d' & ; \\ e : \quad f \quad ; & e + f - a - b - c, c + c' + d' & ; \end{matrix} 1, 1\right]$$
$$= \Gamma\left[\begin{matrix} f, e + f - a - b - c \\ f - c, e + f - a - b \end{matrix}\right] {}_4F_3\left[\begin{matrix} e - a, e - b, c + c', c + d' \\ e, e + f - a - b, c + c' + d' \end{matrix} 1\right] \tag{7.123}$$

Let $\Re(f - a - b - n) > 0$ and n be non-negative integers, then

$$F^{0:3;4}_{1:1;2}\left[\begin{matrix} - : & a, b, e + n & ; -n, e - a, b', c' ; \\ e : & f - a - b - n, 1 + b + b' + c' - f & ; \end{matrix} 1, 1\right]$$
$$= \frac{(e - a, 1 + b + b' - f, 1 + b + c' - f)_n}{(e, 1 + a + b - f, 1 + b + b' + c' - f)_n} \Gamma\left[\begin{matrix} f, f - a - b \\ f - a, f - b \end{matrix}\right] \tag{7.124}$$
$$\times {}_4F_3\left[\begin{matrix} -n, a, f - e - n, f - b - b' - c' - n \\ 1 + a - e - n, f - b - b' - n, f - b - c' - n \end{matrix} 1\right].$$

Let $\Re(f - a - b - n) > 0$ and n be non-negative integers, then

$$F^{0:3;4}_{1:1;2}\left[\begin{matrix} - : a, b, e + n ; & -n, e - a, e - b, c' ; \\ e : \quad f \quad ; & f - a - b - n, f' \end{matrix} 1, 1\right]$$
$$= \frac{(1 + e - f)_n}{(1 + a + b - f)_n} \Gamma\left[\begin{matrix} f, f - a - b \\ f - a, f - b \end{matrix}\right] \tag{7.125}$$
$$\times {}_4F_3(-n, e - a, e - b, f' - c'; e, 1 + e - f, f'; 1).$$

By making further specializations, e.g. $c' = e - c$ in (7.123), we can obtain transformation formulas reducing the double series into ${}_3F_2(1)$.

7.9 Transformation formulas for $F^{1:2;2}_{0:2;2}$

Consider the Kampé de Fériet series

$$F^{1:2;2}_{0:2;2}\left[\begin{matrix} e : a, b ; a', b' \\ - : c, d ; c', d' \end{matrix} 1, 1\right] = \sum_{k,\ell=0}^{\infty} \frac{(e)_{k+\ell}(a, b)_k(a', b')_\ell}{(c, d)_k(c', d')_\ell k!\ell!}. \tag{7.126}$$

This series is terminating if one of the following conditions is satisfied:

- e is a negative integer,
- a or b, and a' or b' are negative integers.

It is also convergent when it is terminating in one summation and convergent in the other, i.e. if:

- a is a negative integer and $\Re(c' + d' - a' - b' - e + a) > 0$; or
- a' is a negative integer and $\Re(c + d - a - b - e + a') > 0$.

In this section, we derive a transformation formula relating a $F^{1:2;2}_{0:2;2}(1, 1)$ series to a $F^{0:3;3}_{1:1;1}(1, 1)$ series. Then (7.120)–(7.122) can be used to find the formulas expressing a $F^{1:2;2}_{0:2;2}(1, 1)$ series into a ${}_4F_3(1)$ series.

Proposition. Let n be a non-negative integer. If $\Re(e' - a' - b' - n) > 0$, then

$$F^{0:3;3}_{1:1;1}\left[\begin{matrix} - & : & -n, a, b & ; & d+n, a', b' & ; \\ d & : & e & ; & e' & ; \end{matrix} 1, 1\right] = \frac{(e-a, b)_n}{(d, e)_n}$$
$$\times F^{1:2;2}_{0:2;2}\left[\begin{matrix} d-b & : & -n, 1-e-n & ; & a', b' & ; \\ - & : & 1+a-e-n, 1-b-n & ; & e', d-b & ; \end{matrix} 1, 1\right] \tag{7.127}$$

.

Proof. The left-hand side of (7.127) can be written as

$$\sum_{\ell=0}^{\infty} \frac{(d+n, a', b')_\ell}{\ell!(d, e')_\ell} {}_3F_2(-n, a, b; d+\ell, e; 1). \tag{7.128}$$

For the terminating ${}_3F_2$, one can apply a transformation from Whipple's list between $Fp(0; 4, 5)$ and $Fp(1; 2, 4)$ (cf ch 3, Bailey, see also, equation (XVI), Srinivasa Rao *et al* 1992)

$$ {}_3F_2(-n, a, -b; d+\ell, e; 1) = \frac{(b, e-a)_n}{(d+\ell, e)_n}$$
$$\times {}_3F_2(-n, d+\ell-b, 1-e-n; 1+a-e-n, 1-b-n; 1). \tag{7.129}$$

Writing $(d+n)_\ell$ as $(d)_\ell(d+\ell)_n/(d)_n$, (7.128) reduces to

$$\sum_{\ell=0}^{\infty} \frac{(a', b')_\ell}{\ell!(e')_\ell} \frac{(b, e-a)_n}{(d, e)_n} {}_3F_2$$
$$(-n, d+\ell-b, 1-e-n; 1+a-e-n, 1-b-n; 1), \tag{7.130}$$

and using $(d-b+\ell)_k = (d-b)_{k+\ell}/(d-b)_\ell$, this can be written in terms of $F^{1:2;2}_{0:2;2}$ of (7.123). QED

Corollary. For $\Re(1 + e - a' - b' - e - n) > 0$:

$$F^{1:2;2}_{0:2;2}\left[\begin{matrix} e & : & a, -n & ; & a', b' & ; \\ - & : & c, d & ; & 1+e-c, e & ; \end{matrix} 1, 1\right]$$
$$= \frac{(d-a)_n}{(d)_n}\Gamma\left[\begin{matrix} 1+e-c, 1+e-c-a'-b' \\ 1+e-c-a', 1+e-c-b' \end{matrix}\right] \tag{7.131}$$
$$\times {}_4F_3(-n, a, c-e+a', c-e+b'; c, 1+a-d-n, c-e+a'+b'; 1).$$

For $\Re(c' - a' - b' - n) > 0$:

$$F^{1:2;2}_{0:2;2}\left[\begin{matrix} e & : & a, -n & ; & a', b' \\ - & : & c, e & ; & c', e \end{matrix} \quad 1, 1\right] = \frac{(c-a)_n}{(d)_n}\Gamma\left[\begin{matrix} c', c'-a'-b' \\ c'-a', c'-b' \end{matrix}\right] \tag{7.132}$$
$$\times\, {}_4F_3(-n, a, a', b'; e, 1+a-c-n, 1+a'-c'+b'; 1).$$

For $\Re(b+c-a-e+c'-a'-1) > 0$:

$$F^{1:2;2}_{0:2;2}\left[\begin{matrix} e & : & a, -n & ; & 1+a+e-n-a-b-c; \\ - & : & b, c & ; & c', e; \end{matrix} \quad 1, 1\right]$$
$$= \frac{(d-a)_n}{(d)_n}\Gamma\left[\begin{matrix} 1+e-c, 1+e-c-a'-b' \\ 1+e-c-a', 1+e-c-b' \end{matrix}\right] \tag{7.133}$$
$$\times\, {}_4F_3(-n, b-a, c-a, b+c-a-e+c'-a'-1;$$
$$1-a-n, b+c-a-e, b+c-a-e+c'-1; 1).$$

Proof. Put $e' = d+e+n-a-b$, $e = 1+a+b-d-n$ and $a = d-a'$ in equation (7.127) and apply (7.120), (7.121), and (7.122).

Proposition. For n, a non-negative integer there holds

$$F^{1:2;2}_{0:2;2}\left[\begin{matrix} -n & : & a, b & ; & a', 1+a+b-c-d-n; \\ - & : & c, d & ; & c', 1-d-n; \end{matrix} \quad 1, 1\right]$$
$$= \frac{(c+d-a-b, c'-a-)_n}{(d, c')_n} \tag{7.134}$$
$$\times\, {}_4F_3(-n, c-a, c-b, 1-c'-n; c, c+d-a-b, a+a'-c'-n; 1).$$

Proof. Denoting the left-hand side of (7.134) by L, it can be written as

$$L = \sum_{\ell=0}^{n} \frac{(-n, a', 1+a+b-c-d-n)_\ell}{\ell!(c', 1-d-n)_l}\, {}_3F_2(-n+l, a, b; c, d; 1). \tag{7.135}$$

For ${}_3F_2$, applying a transformation from Whipple's list (Bailey (1935), ch 3) between $F_p(0; 4,5)$ and $F_p(1; 2,3)$, leads to:

$${}_3F_2(-n+l, a, b; c, d; 1) = \frac{(1-b+\ell-n)_{n-\ell}}{(c, d)_{n-\ell}}$$
$$\times\,\Gamma\left[\begin{matrix} 1+a-d, 1+a-c \\ 1+a-b, 1+a+b-c-d-n+\ell \end{matrix}\right] \tag{7.136}$$
$$\times\, {}_3F_2(1-b, c-b, d-b; 1+a-b, 1-b-n+l; 1).$$

Substituting (7.136) in (7.135) yields, after some simplifications,

$$L = \frac{(1-b-n)_n}{(c,d)_n}\Gamma\left[\begin{matrix}1-c+a,\ 1-d+a\\ 1+a-b,\ 1+a+b-c-d-n\end{matrix}\right] \times F^{0:3;3}_{1:1;1}\left[\begin{matrix}- & : & -n, a', 1-c-n & ; & 1-b, c-b, d-b\\ 1-b-n & : & c' & ; & 1+a-b\end{matrix}\ 1, 1\right]. \tag{7.137}$$

Next, we apply (7.135), and obtain

$$\begin{aligned}L = &\frac{(1+a-c-n, c-b)_n}{(c,d)_n}\\ &\times {}_4F_3(-n, 1-c-n, 1+a+b-c-d-n,\\ &c'-a'; c', 1+a-c-n, 1+b-c-n; 1).\end{aligned} \tag{7.138}$$

Finally, performing a reversal of the series on ${}_4F_3$ leads to (7.134).

7.10 Some summation formulas

Here, some limiting cases of the above transformation formulas are considered. In this section, m and n always denote non-negative integers. Putting $f' = c'$ in (7.125) leads directly to:

$$\begin{aligned}&F^{0:3;3}_{1:1;1}\left[\begin{matrix}- & : & a, b, e+n & ; & -n, e-a, e-b & ;\\ e & : & f & ; & f-a-b-n & ;\end{matrix}\ 1, 1\right]\\ &= \frac{(1+e-f)_n}{(1+a+b-f)_n}\Gamma\left[\begin{matrix}f, f-a-b\\ f-a, f-b\end{matrix}\right],\end{aligned} \tag{7.139}$$

where $\Re(f-a-b-n) > 0$. Substituting $b' = 1+a+e-n-c-d$ and $a' = d+n-1$ in (7.131), ${}_4F_3(1)$ reduces to ${}_2F_1(1)$, leading to

$$\begin{aligned}&F^{1:2;2}_{0:2;2}\left[\begin{matrix}e & : & a, -n & ; & d+n-1, 1+a-e-n-c-d & ;\\ - & : & c, d & ; & 1+e-c, e & ;\end{matrix}\ 1, 1\right]\\ &= \frac{(d-e, c+d-e-1)_n}{(d,c)_n}\Gamma\left[\begin{matrix}1+e-c, 1-a\\ d-a, 2+e-c-d\end{matrix}\right],\end{aligned} \tag{7.140}$$

where $\Re(1+c-a-e-n) > 0$. This can be further specialized to a double series terminating in both variables:

$$\begin{aligned}&F^{1:2;2}_{0:2;2}\left[\begin{matrix}e & : & a, -n & ; & 1+a-c-n, b' & ;\\ - & : & c, e & ; & 2+b'-c-n, e & ;\end{matrix}\ 1, 1\right]\\ &= (-1)^m\frac{(a)_m(c-e-m, c-a-m)_n}{(1+e-c)_m(c, c-a-e-m)_n}.\end{aligned} \tag{7.141}$$

Consider next (7.132) under the extra conditions $a' = 1 + a - c - n$ and $c' = 2 + b' - c - n$. There comes

$$F^{1:2;2}_{0:2;2}\left[\begin{matrix} e & : & a, -n & ; & 1 + a - c - n, b' & ; \\ - & : & c, e & ; & 2 + b' - c - n, e & ; \end{matrix}\right] = \frac{(c - a, e - b')_n}{(c, e)_n}\Gamma\left[\begin{matrix} 2 + b' - c - n, 1 - a \\ 1 + b' - a, 2 - c - n \end{matrix}\right], \tag{7.142}$$

where $\Re(1 - a - n) > 0$. For both sides terminating, i.e. $b' = -m$, this reduces to Carlitz's identity (Carlitz 1967):

$$F^{1:2;2}_{0:2;2}\left[\begin{matrix} e & : & a, -n & ; & a', -m & ; \\ - & : & c, e & ; & c', e & ; \end{matrix}\right] = \frac{(e)_{m+n}(c - a)_n\,(c' - a')_m}{(c, e)_n\,(c', e)_m}, \tag{7.143}$$

where the parameters must satisfy $a' = 1 + a - c - n$ and $a = 1 + a' - c' - m$.

Consider again (7.132), but now with $a' = 1 + a - c - n$ and $c' = 1 + e - c$; one gets

$$F^{1:2;2}_{0:2;2}\left[\begin{matrix} e & : & a, -n & ; & a', -m & ; \\ - & : & c, e & ; & c, e & ; \end{matrix}\right] = \frac{(c - a, e - b')_n}{(c, e)_n}\Gamma\left[\begin{matrix} 1 + e - c, e - a - b' \\ 1 + e - c - b', e - a \end{matrix}\right], \tag{7.144}$$

for $\Re(e - a - b') > 0$. For $b' = -m$, both sides are terminating, and one obtains

$$F^{1:2;2}_{0:2;2}\left[\begin{matrix} e & : & a, -n & ; & 1 + a - c - n, -m & ; \\ - & : & c, e & ; & 1 + e - c, e & ; \end{matrix} 1, 1\right] = \frac{(e)_{m+n}(c - a)_n\,(e - a)_m}{(c, e)_n\,(1 + e - c, e)_m}. \tag{7.145}$$

Finally, an interesting formula is obtained from (7.134) by choosing $a' = -a$ and $c' = 1 - c - n$:

$$F^{1:2;2}_{0:2;2}\left[\begin{matrix} -n & : & a, b & ; & -a, 1 + a + b - c - d - n & ; \\ - & : & c, d & ; & 1 - c - n, 1 - d - n & ; \end{matrix}\right] = \frac{(c - a, d - a)_n}{(c, d)_n}. \tag{7.146}$$

The transformation formulas presented here were obtained in the context of a study of the series expressions for symmetries of the 9-j angular momentum recoupling coefficient. In particular, the five distinct types of doubly stretched 9-j coefficients have been considered, and a complete classification of the series related to one of the five types of doubly stretched 9-j coefficients was given in Van der Jeugt *et al* (1994). The classification of the remaining types was completed in the PhD thesis by S N Pitre (University of Ghent 1996).

For a single series appearing in this framework, no new summation or transformation formulas were obtained: they turned out to be one of the classical results—the summation theorems of Vandermonde, Saalschütz, or Karlsson–Minton—or the ${}_3F_2(1)$ summation formulas of Thomae, Weber–Erdélyi; or the ${}_3F_2$ transformation formula of Whipple and Bailey; or the Saalschützian ${}_4F_3$ transformation formula.

References

Alisauskas S J and Jucys A P 1971 *J. Math. Phys.* **12** 594 Erratum 1972 *J. Math. Phys.* **13** 575.

Appell M P 1925 *Sur Les Fonctions Hypergeometrique de Plusieurs Variables* (Paris: Gauthier-Villars)

Appell P and Kampe de Feriet J 1926 *Fonctions Hypergeometriques et Hyperspheriques* (Pans: Gauthier-Villars)

Bailey W N 1935 *Generalized Hypergeometric Series* Cambridge Tracts in Mathematics and Mathematical Physics 32 (Cambridge: Cambridge University Press)

Bandzaitis A A, Karosienne A and Jucys A P 1964 *Liet. Fiz. Rin.* **4** 457

Biedenharn L C and Louck J D 1981 *Angular Momentum in Quantum Physics: Theory and Application* (Encyclopedia of Mathematics and its Applications vol 8) (Cambridge: Cambridge University Press)

Biedenharn L C and Louck J D 1984 *The Racah Wigner Algebra in Quantum Theory* (Encyclopedia of Mathematics and Its Applications vol 9) (Cambridge: Cambridge Unviersity Press)

Carlitz L 1967a *Mathematica (Catania)* **22** 138

Carlitz L 1967b *Uncertainty: The Life and Science of Werner Heisenberg* (San Francisco, CA: Freeman)

Edmonds A R 1957 *Angular Momentum in Quantum Mechanics* (Princeton, NJ: Princeton University Press)

Exton H 1976 *Multiple Hypergeometric Functions and Applications* (Mathematics and its Applications) (Chichester: Ellis Horwood)

Gasper G and Rahmann M 1991 *Basic Hypergeometric Series* (Cambridge: Cambridge University Press) Also, 1990 Encyclopedia of Mathematics and its Applications vol 35 (Cambridge: Cambridge University Press)

Hai N T, Marichev O I and Srivastava H M 1992 *The Double Mellin–Barnes Type Integrals and their Applications to Convolution Theory* Series on Soviet and East European Mathematics vol 6 ed N T Hai and S B Yakubovich (Singapore: World Scientific)

Horn J 1931 corrected by Borngässer, Ludwig, Über hypergeometrische funkionen zweier Veränderlichen, *Dissertation* Darmstadt.

Jain R N 1966 *Mathematica (Catania)* **21** 300

Jucys A P and Bandzitis A A 1977 The Theory of Angular Momentum in Quantum Mechanics (Vilnius: Mintis) (original in Russian, 1965)

Kampé de Fériet 1923 Les Fonctions hypergeometrique d'ordere superiur a deux variables *Comptes rendus* CLXXIII

Karlsson P W 1971 *J. Math. Phys.* **12** 270

Lakshminarayanan V and Varadharajan L S 2015 Special Functions for Optical Science and Engineering Vol 2 (Bellingham, WA: SPIE Press)
Lathrop C and Stemkoski L 2007 Parallels in the Work of Leonhard Euler and Thomas Clausen *Euler at 300* ed E Bradley Robert, A D'Antonio Lawrence and S C Edward (Washington, DC: Math Association America), pp 217–25
Lauricella G 1893 Sulle funzioni ipergeometriche a più variabili R. C. Mate *Palermo* **7** 111–58
Miller A R and Srivastava H M 2010 *Integr. Transf. Spec. Funct.* **21** 603–12
Minton B M 1970 *Math. Phys.* **11** 3061
Pitre S N 1996 *PhD Thesis* University of Ghent, Ghent, Belgium.
Pitre S N and Van der Jeugt J 1996 *J. Math. Anal. Appl* **202** 21
Saran S 1954 Hypergeometric Functions of Three Variables *Ganita.* **5** 77
Sharp R T 1967 *Nucl. Phys.* B **2** 222–28
Sharp R T and von Baeyer H 1966 *J. Math. Phys.* **7** 1105
Sharp R T and von Baeyer H C 1967 *Nucl. Phys. A* **95** 222–28
Slater L J 1966 *Generalized Hypergeometric Functions* (Cambridge: Cambridge University Press)
Slater L J 1960 *Confluent Hypergeometric Functions* (Cambridge: Cambridge University Press)
Srinivasa Rao K and Rajeswari V 1993 *Selected Topics in Quantum Theory of Angular Momeutm* (New York: Springer)
Srinivasa Rao K and Rajeswari V 1989 *J. Math. Phys.* **30** 1016
Srinivasa Rao K and Van der Jeugt J 1994 *J. Phys. A: Math. Gen.* **27** 3083
Srinivasa Rao K, Van der Jeugt J, Raynal J, Jagannathan R and Rajeswari V 1992 *J. Phys. A: Math. Gen.* **25** 861–76
Srivastava H M 1964 *Ganita Bharati* **1597**
Srivastava H M 1985 *J. Phys. A: Math. Gen.* **18** L227
Srivastava H M and Daoust M 1972 *Math. Nachr.* **53** 151
Srivastava H M and Karlsson Per W 1985 *Multiple Gaussian Hypergeometric Series* (Chichester: Ellis Horwood), p 270
Srivastava H M 1967 *Multiple Gaussian Hypergeometric Series* Ellis Horwood Series in Mathematics and its Applications.
Van der Jeugt J and Pitre S N 1996
van der Jeugt J, Pitre S N and Srinivasa Rao K 1994 *J. Phys. A: Math. Gen.* **27** 5251
Van der Jeugt J, Pitre S N and Srinivasa Rao K 1997 *J. Comput. Appl. Math.* **83** 185–93 See also, 1998 *Special Functions and Differential Equations*, ed K Srinivasa Rao, R Jagannathan, G Vanden Berghe and J Van der Jeugt (New Delhi: Allied Publishers) pp 171–7.
Wigner E P 1940 The unreasonable effectiveness of mathematics in the natural sciences *Commun. Pure Appl. Math.* **13** 1–14
Wu A C T 1972 *J. Math. Phys.* **13** 84
Wu A C T 1973 *J. Math. Phys.* **14** 1222

Chapter 8

Beta integral method and hypergeometric transformations

8.1 Introduction

The simplest special functions are the gamma function and the beta function:

$$\Gamma(n+1) = n! \quad \text{and} \quad B(m, n) = \frac{\Gamma(m)\Gamma(n)}{\Gamma(m+n)}. \tag{8.1}$$

Euler's beta integral evaluation

$$\int_0^1 z^{\alpha-1}(1-s)^{\beta-1}dz = \frac{\Gamma(\alpha)\Gamma(\beta)}{\Gamma(\alpha+\beta)}, \quad \Re(\alpha) > 0, \Re(\beta) > 0 \tag{8.2}$$

is the source for many hypergeometric series/function identities. Luke (1969, equation (11), section 3.6) provides the integral representation for the generalized hypergeometric function as

$$\begin{aligned} {}_{p+1}F_p(\alpha, \alpha_1, \alpha_2, \ldots, \alpha_p; \gamma, \beta_1, \beta_2, \ldots, \beta_1 p-1; t) &= \frac{\gamma(\gamma)}{\gamma(\alpha)\Gamma(\gamma-\alpha)} \\ \times \int_0^1 z^{\alpha-1}(1-z)^{\gamma-\alpha-1} {}_pF_{p-1}(\alpha_1, \alpha_2, \ldots, \alpha_p; \beta_1, \beta_2, \ldots, \beta_{p-1};\ zt)\, dz, \end{aligned} \tag{8.3}$$

where the generalized hypergeoemtric function has the series representation

$$\begin{aligned} &{}_{p+1}F_p(\alpha, \alpha_1, \alpha_2, \ldots, \alpha_p; \gamma, \beta_1, \beta_2, \ldots, \beta_1 p-1; t) \\ &= \sum_{k=0}^{\infty} \frac{(\alpha)_k(\alpha_1+1)_k \cdots (\alpha_p+1)_k}{(\gamma)_k(\beta_1+1)_k \cdots (\beta_{p-1}+1)_k} \frac{(zt)^k}{k!}. \end{aligned} \tag{8.4}$$

We insert (8.4) in (8.3), interchange the integration and summation, and then use (8.2) to evaluate the integral inside the summation to prove (8.3).

doi:10.1088/978-0-7503-1496-1ch8

Krattenthaler and Srinivasa Rao (2003) proposed a *beta integral method* to derive new hypergoemetric series identities from known old ones, using the beta integral evaluation.

8.2 Extensions of Euler's integral for $_2F_1(a, b; c; z)$

Erdélyi (1939) showed that Euler's integral

$$
{}_2F_1(a, b; c; z) = \frac{\Gamma(c)}{\Gamma(b)\Gamma(c-b)} \int_0^1 t^{b-1}(1-t)^{b-1}(1-zt)^{c-b-1}\,dt, \qquad z \neq 1, \quad |arg(1-z)| < \pi, \quad \Re(a) > 0, \Re(b) > 0, \tag{8.5}
$$

and its extension by Bateman (Erdélyi *et al* 1939)

$$
{}_2F_1(a, b; c; z) = \frac{\Gamma(c)}{\Gamma(m)\Gamma(c-m)} \int_0^1 t^{m-1}(1-t)^{c-m-1}{}_2F_1(a, b; m; zt)\,dt, \qquad \Re(c) > \Re(m) > 0, \tag{8.6}
$$

have extensions of the forms

$$
\begin{aligned}
{}_2F_1(a, b; c; z) &= \frac{\Gamma(c)}{\Gamma(m)\Gamma(c-m)} \int_0^1 t^{m-1}(1-t)^{c-m-1}(1-zt)^{\ell-a-b} \\
&\times {}_2F_1(\ell-a, \ell-b; m; zt){}_2F_1\left(a+b-\ell, \ell-m; c-m; \frac{(1-t)z}{1-zt}\right) dt, \\
&\Re(c) > \Re(m) > 0,
\end{aligned} \tag{8.7}
$$

$$
\begin{aligned}
{}_2F_1(a, b; c; z) &= \frac{\Gamma(c)}{\Gamma(m)\Gamma(c-m)} \int_0^1 t^{m-1}(1-t)^{c-m-1}(1-zt)^{-a'} \\
&\times {}_2F_1(a-a', b; m; zt){}_2F_1\left(a', b-m; c-m; \frac{(1-t)z}{1-zt}\right) dt, \\
&\Re(c) > \Re(m) > 0,
\end{aligned} \tag{8.8}
$$

and

$$
\begin{aligned}
{}_2F_1(a, b; c; z) &= \frac{\Gamma(c)\Gamma(m)}{\Gamma(\ell)\Gamma(n)\Gamma(c+m-\ell-n)} \int_0^1 t^{m-1}(1-t)^{c+m-\ell-n-1}(1-tz)^{-a'} \\
&\times {}_2F_1(m-\ell, c-\ell; c+m-\ell-n; 1-t){}_3F_2(a, b, m; \ell, n; zt)\,dt, \\
&\Re(\ell, n, c+m-\ell-n) > 0.
\end{aligned} \tag{8.9}
$$

Where, in (8.6) $m = b$, in

$$
{}_2F_1(a, b; b; zt) = {}_1F_0(a; -; zt) = (1-zt)^{-a} \tag{8.10}
$$

it becomes (8.5). Also, when $\ell = a + b$, $a' = 0$, $\ell = \mathrm{m}$, equations (8.7)–(8.9) become Bateman's (8.6). Erdélyi considered special cases as well as the cases that arise in the limit $b \to \infty$, which reduces the Gauss second order ordinary differential equation into the differential equation for the confluent hypergeometric series by applying transformation formulas to the hypergeometric functions in the integrand. Gasper (1975) derived the integral representations (called the Dirichlet–Mehler-type) for the Jacobi polynomials and for the generalized Legendre functions. He proved the positivity of certain sums of generalized Legendre functions, derived the discrete analogue of Bateman's extension (8.6), and derived the discrete analogue of (8.7). This work was later extended by Gasper (2000) for (8.7) and (8.8). Using the series manipulation technique and the classical summation theorems, Joshi and Vyas (2003) gave an alternative proof for the Bateman extension of the Erdélyi integrals (8.6)–(8.8). They also conjectured and proved new integrals of the Erdélyi type for certain ${}_{p+1}F_p(z)$.

The research work on Bateman's equation (8.6) has been extended by Srivastava and Karlsson (1985), also by Appell, Kampé de Feriet, Lauricella; plus other multiple hypergoemetric series' using fractional calculus.

8.3 The beta integral method

Krattenthaler and Srinivasa Rao (2003) proposed what they call the *beta integral method* and exploit the same to derive new hypergeometric identities from existing old ones. The method itself is folklore in the literature on the hypergeometric series (see Andrews *et al* 1999), though it is not given a name for the simple technique. Although appearances of this method can be found sporadically—for instance in Andrews, Askey, Ranjan Roy (1999; chapter 3, exercises 5, 14 and 16)—it has perhaps not been exploited systematically. That is, until in this work of Krattenthaler and Srinivasa Rao (2003)—probably because of the effort it takes to do the computations. Here, we reproduce the *completely automatic* application of the beta integral method.

To convey the spirit of the method (perhaps not the first one), Weber and Erdelyi (1952, section 3) start with the well-known transformation formula in Slater (1966, equation (1.8.10)):

$$ {}_2F_1(a, -n; c; z) = \frac{(c-a)_n}{(c)_n} {}_2F_1(a, -n; a-c-n+1; 1-z), \tag{8.11} $$

multiply both sides by

$$ z^{b-1}(1-z)^{d-b-1}, \tag{8.12} $$

and then integrate both sides with respect to z, $0 \leqslant z \leqslant 1$. They interchange the integration and summation on both sides, then use (8.2), and, finally, convert the result back to the hypergeometric notation to get the transformation formula:

$$ {}_3F_2(a, b, -n; c, d; 1) = \frac{\Gamma(c-a)_n}{(c)_n} {}_3F_2(a, d-b, -n; 1+a-c-n, d; 1), \tag{8.13} $$

where n is a non-negative integer. In this case, we obtained an already known transformation formula, namely one of the Thomae transformation formulas, in (8.13).

Krattenthaler and Srinivasa Rao (2003) applied this method to a large variety of identities and completely automated the method with the help of the Mathematica program, HYP, created by Krattenthaler (1995). The beta integral method is similar in spirit to the idea of *dual identities* in Wilf–Zeilberger theory (1992), and the idea of *parameter augmentation*, because of the introduction of one additional parameter.

This algorithm has also been applied to transformation formulas for products of hypergeometric functions known in the literature. As a result, formulae for the Kampé de Friet series of unit arguments, which transform them into the single-sum hypergeometric series, are obtained. Again, some of these results have been previously obtained and are known; but others are new and interesting.

8.4 The algorithm

The derivation of (8.13) from (8.11) involves the following basic steps:

- (i) convert the hypergeometric series on both sides of a given transformation into sums;
- (ii) multiply both sides of the equation by the factors $z^{b-1}(1-z)^{d-b-1}$;
- (iii) integrate each term by term with respect to z for $0 \leqslant z \leqslant 1$;
- (iv) interchange integration and summation;
- (v) use the beta integral to evaluate the integrals inside the summations;
- (vi) convert the sums back into the hypergeometric function notation.

Christian Krattenthaler[1] created two Mathematica based software packages *HYP* and *HYP-q*. We reproduce below—from the home page of Krattenthaler—the precise details of the capabilities of this software, which is a boon to mathematicians.

The Mathematica package HYP

HYP is a package, written in Mathematica, for the manipulation and identification of the binomial and hypergeometric series and identities.

This package provides tools for:

- (A) manipulating factorial expressions;
- (B) transforming binomial sums into hypergeometric notation;
- (C) summing hypergeometric series;
- (D) transforming hypergeometric series;
- (E) applying contiguous relations;
- (F) doing formal limits of hypergeometric expressions;
- (G) transforming hypergeometric expressions into TeX-code;
- (H) using the Gosper and Zeilberger algorithms.

(For item (H) you need Peter Paule and Markus Schorn's Mathematica implementation of the Gosper and Zeilberger algorithms, which is available at: https://www.risc.jku.at/home/ppaule/software.shtml. You should also get Marko Petkovsek's (see Petkovsek and Salvy 1993) (*Marko:Petkovsek@fmf, uni–lj.si*) Mathematica

[1] www.mat.univie.ac.at/kratt/hyphypq/hyp.html.

program, Hyper, which finds all hypergeometric term solutions of a linear recurrence with polynomial coefficients[2].)

The package comes with complete documentation, a tutorial, and an installation guide. It works with any version of Mathematica, except that Gosper's and Zeilberger's algorithms require Version 2.0 or higher.

Retrieve the files directly, with the following to be placed after 'www.':

hyp.m, the basic input;
contig.m, the input for contiguous relations;
summatio.m, the input for summations in form of rules;
summatio.mgl, the input for summations in form of equations;
transfor.m, the input for transformations in form of rules;
transfor.mgl, the input for transformations in form of equations;
transfor.mli, the input for lists of transformations;
output, a file that makes a nicer screen output;
read.me, the installation guide;
hypm.tex, the AmS-TeX file for the documentation (453 K);
hypm.dvi, the dvi-file for the documentation (590 K);
hypm.ps.gz, the gzipped postscript-file for the documentation (206 K);
hypm.pdf, the PDF file for the documentation (552 K);
hyp-hypq.tex, the AmS-TeX file for a tutorial;
hyp-hypq.dvi, the dvi-file for a tutorial;
hyp-hypq.ps.gz, the gzipped postsript-file for a tutorial (45 K);
hyp-hypq.pdf, the PDF file for a tutorial;
'HYP and HYPQ', a short article that appeared in J. Symbol. Comput. 20 (1995), 737–44.

Read the file read.chg for the differences between the current version and older versions. If you want to be informed about updates for these packages, or if you have comments, suggestions, or problems, please send an e-mail to kratt@ap.univie.ac.at.

There is also a 'q-analogue': a package HYPQ that allows you to manipulate and identify the basic hypergeometric series. This was taken from Prof Dr Christian Krattenthaler's home page.

In the next section, we present some of the results obtained using the HYP software package by Krattenthaler and Srinivasa Rao (2003).

8.5 New single sum hypergeometric identities from old ones

The algorithm of section 8.4 above is applied systematically to known transformations of the hypergeometric series. It was already shown that the result (8.13) for ${}_3F_2(1)$ is due to the existence of the ${}_2F_1$ transformation in (8.11) using

[2] https://www.fmf.uni-lj.si/petkovsek/sofware.htm.

the beta integral method. The algorithm is stated explicitly in section 8.4. The transformation formula (Slater 1966, (1.7.1.3)):

$$ {}_2F_1(a, b; c; z) = (1 - z)^{-a} {}_2F_1\left(a, c - b; c; -\frac{z}{1 - z}\right) \tag{8.14} $$

will also result in one of the 18 terminating ${}_3F_2(1)$ transformations—given in chapter 4, p 86—when one of the numerator parameters is a negative integer.

1. Let us start with the quadratic transformation formula (3.2) given in Rahman and Verma (1993):

$$ \begin{aligned} {}_2F_1(a, b; 1 + a - b; -z) &= (1 - z)^{-a} \\ & {}_2F_1\left(\frac{a}{2}, \frac{1}{2} + \frac{a}{2}; 1 + a - b; -\frac{4z}{(1 - z)^2}\right) \end{aligned} \tag{8.15} $$

   and assume that a is a negative integer to terminate the ${}_2F_1$ series. Then we get

$$ \begin{aligned} {}_3F_2(a, b, d; a + a - b, e; -1) &= \frac{\Gamma(e, e - a - d)}{\Gamma(e - a, e - d)} \\ & {}_4F_3\left(\frac{1}{2} + \frac{a}{2}, \frac{a}{2}, d, 1 + a - e; a + a - b, \right. \\ & \left. \frac{1}{2} + \frac{a}{2} + \frac{d}{2} - \frac{e}{2}, 1 + \frac{a}{2} + \frac{d}{2} - \frac{e}{2}\right). \end{aligned} \tag{8.16} $$

   This transformation is valid, relating a nearly poised ${}_3F_2(-1)$ to ${}_4F_3(1)$; provided a is a non-positive integer. However, by a standard polynomial trick, it can be shown that it is also true if a is arbitrary, but d is a non-positive integer: let d be a non-positive integer. By multiplying both sides of (8.16) by

$$ (e - a)_d(1 + a - b)_d(1 + a + d - e)_{2d}, $$

   both sides become polynomials in a of the degree at most $5d$. These two polynomials agree for all of the non-positive a, since we know from (8.16) that it is true for the non-positive a. There are infinitely many values of a, whence the polynomials must be identical.

2. If we start with the transformation formula (5.10) of Krattenthaler in http://www.mat.univie.ac.at/~kratt:

$$ {}_2F_1\left(a, \frac{1}{2} + q; \frac{1}{2} + b; z^2\right) = (1 - z)^{-2a} {}_2F_1\left(2a, b; 2b; \frac{2z}{z - 1}\right) \tag{8.17} $$

   and assume that a is a negative integer, we then obtain

$$
\begin{aligned}
&{}_4F_3\left(a, \frac{1}{2}+a, \frac{1}{2}+\frac{d}{2}, \frac{d}{2}; \frac{1}{2}+b, \frac{1}{2}+\frac{e}{2}, \frac{e}{2}; 1\right) \\
&= \frac{\Gamma(e, e-2a-d)}{\Gamma(e-2a, e-d)}\,{}_3F_2(2a, b, d; 2b, 1+2a+d-e; 2).
\end{aligned} \tag{8.18}
$$

This identity is true if a is a non-positive integer, or if d is a non-positive integer; the latter can be attributed to the same arguments stated above for item 1. It can be found in the literature as a special case of a more general transformation for the basic hypergeometric series (see Gasper and Rahman 1990, exercise 3.4, $q \to 1$, reverse; available as T3235 in HYP in the manual by Kratternthaler for HYP).

3. If we start with the quadratic transformation formula of Gauss (Bailey 1935, example 4(iii), p 97), with $\alpha \to a/2$, $\beta \to b/2$, $x \to z$

$$
{}_2F_1\left(a, b; \frac{1}{2}+\frac{a}{2}+\frac{b}{2}; z\right) = {}_2F_1\left(\frac{a}{2}, \frac{b}{2}; \frac{1}{2}+\frac{a}{2}+\frac{b}{2}; 4(1-z)z\right), \tag{8.19}
$$

and assume that a is a non-positive integer, then we obtain

$$
\begin{aligned}
&{}_3F_2\left(a, b, d; \frac{1}{2}+\frac{a}{2}+\frac{b}{2}, e; 1\right) \\
&= {}_4F_3\left(\frac{a}{2}, \frac{b}{2}, d, -d+e; \frac{1}{2}+\frac{a}{2}+\frac{b}{2}, \frac{1}{2}+\frac{e}{2}; 1\right).
\end{aligned} \tag{8.20}
$$

This identity is true provided both the hypergeometric series' terminate. It is the main theorem in Slater (1966) and also in Gaper and Rahman (1990, (3.10.13); appendix (III.21)). This same result (8.20) will be obtained when $z \to 1-z$ in (8.19).

4. If we start with the quadratic transformation of Rahman and Verma (1993, (5.12) reversed):

$$
{}_2F_1(a, b; 2b; z) = (1-z)^{-a/2}\,{}_2F_1\left(\frac{a}{2}, -\frac{a}{2}+b; \frac{1}{2}+b; \frac{z^2}{4(z-1)}\right) \tag{8.21}
$$

and assume that a is an non-positive integer, then

$$
\begin{aligned}
{}_3F_2(a, b, d; 2b, e; 1) &= \frac{\Gamma(c, e-\frac{a}{2}-d)}{\Gamma(e-\frac{a}{2}, e-d)} \\
&\times {}_4F_3\left(\frac{a}{2}, -\frac{a}{2}+b, \frac{1}{2}+\frac{d}{2}, \frac{d}{2}; \frac{1}{2}+b, 1+\frac{a}{2}+d-e, -\frac{a}{2}+e; 1\right)
\end{aligned} \tag{8.22}
$$

provided a is an even non-positive, or d is any non-positive integer. This identity can also be obtained in a different (but more complicated) way. In the transformation formula listed as T4391 in HYP (Krattenthaler (1995), equation (3.5.7) from Gasper and Rahman (1990) with $q \to 1$), let $e \to \infty$. Then, on the right-hand side, the second term vanishes, while the first is a very-well-poised ${}_7F_6$-series for Whipple's ${}_7F_6$ to ${}_4F_3$ transformation (see Slater 1966, (2.4.1.1), reversed).

5. The transformation formula (Rahman and Verma 1993, (3.31), reversed)

$$ {}_2F_1(a, 1-a; c; z) = (1-z)^{c-1}{}_2F_1\left(-\frac{a}{2}+\frac{c}{2}, -\frac{1}{2}+\frac{a}{2}+\frac{c}{2}; c; 4(1-z)z\right) \tag{8.23} $$

results in

$$ \begin{aligned} {}_3F_2(1-a, a, d; c, e; 1) &= \frac{\Gamma(e, c-d+e-1)}{\Gamma(c+e-1, e-d)} \\ \times\, {}_4F_3\Bigg(&-\frac{a}{2}+\frac{c}{2}, -\frac{1}{2}+\frac{a}{2}+\frac{c}{2}, d, c-d+e-1; \\ &c, -\frac{1}{2}+\frac{c}{2}+\frac{e}{2}, \frac{c}{2}+\frac{e}{2}; 1\Bigg) \end{aligned} \tag{8.24} $$

provided both hypergeometric series' terminate.

6. If we start with the transformation formula in Bailey (1928, (4.10), with $x \to -z/2$):

$$ {}_2F_1\left(a, b; 2b; -\frac{4z}{(1-z)^2}\right) = (1-z)^{2a}{}_2F_1\left(a, \frac{1}{2}+a-b; \frac{1}{2}+b; z^2\right), \tag{8.25} $$

which is a combination of Rahman and Verma (1993, (5.10) and (6.2) reversed), and assume that a is a non-positive integer, then we obtain

$$ \begin{aligned} &{}_4F_3\left(a, b, d, e; 2b, \frac{d}{2}+\frac{e}{2}, \frac{1}{2}+\frac{d}{2}+\frac{e}{2}; 1\right) \\ &= \frac{\Gamma(1-e, 1+2a-d-e)}{\Gamma(1+2a-e, 1-d-e)} \\ &{}_4F_3\Bigg(a, \frac{1}{2}+a-b, \frac{1}{2}+\frac{d}{2}, \frac{d}{2}; \frac{1}{2}+b, \\ &\frac{1}{2}+a-\frac{e}{2}, 1+a-\frac{e}{2}; 1\Bigg), \end{aligned} \tag{8.26} $$

provided a or d is a non-positive integer.

7. If we start with the transformation formula in Bailey (1928, (4.22), with $\alpha \to a$, $\beta \to b$, $x \to z$) and Rahman and Verma (1993, (5.12) with $z \to z/(2-z)$)

$$ {}_2F_1\left(a, b; \frac{1}{2}+a+b; \frac{z^2}{4(z-1)}\right) = (1-z)^a{}_2F_1(2a, a+b; 2a+2b; z), \tag{8.27} $$

and assume that a is a non-positive integer, then we obtain

$$
\begin{aligned}
&{}_4F_3\left(a, b, \frac{1}{2} + \frac{d}{2}, \frac{d}{2}; \frac{1}{2} + a + b, 1 + d - e, e; 1\right) \\
&= \frac{\Gamma(e, a - d - e)}{\Gamma(a + e, e - d)}\, {}_3F_2(2a, a + b, d; 2a + 2b, a + e; 1),
\end{aligned}
\tag{8.28}
$$

provided a or d is a non-positive integer.

8. The transformation formula in Rahman and Verma (1993, equation (3.31)) with $z \to 4z/(1 - z)$, $a \to a + c + \frac{1}{2}$, $c \to b$):

$$
\begin{aligned}
&{}_2F_1\left(a, b; \frac{1}{2} + a + b; 4(1 - z)z\right) \\
&= (1 - z)^{1/2-a-b}\, {}_2F_1\left(\frac{1}{2} + a - b, \frac{1}{2} - a + b; \frac{1}{2} + a + b; z\right),
\end{aligned}
\tag{8.29}
$$

and assume that a is a non-positive integer, then we obtain

$$
\begin{aligned}
&{}_4F_3\left(a, b, d, e - d; \frac{1}{2} + a + b, \frac{1}{2} + \frac{e}{2}, \frac{e}{2}; 1\right) \\
&= \frac{\Gamma\left(e, \frac{1}{2} - a - b - d + e\right)}{\Gamma\left(\frac{1}{2} - a - b + e, e - d\right)} \\
&\times {}_3F_2\left(\frac{1}{2} + a - b, \frac{1}{2} - a + b, d; \frac{1}{2} + a + b, \frac{1}{2} - a - b + e; 1\right),
\end{aligned}
\tag{8.30}
$$

provided both hypergeometric series' terminate. This transformation can also be obtained by combining the ${}_4F_3$ to ${}_3F_2$ transformation from V N Singh (1959) that occurred already in item three with the ${}_3F_2$ transformation:

$$
\begin{aligned}
{}_3F_2(a, b, c; d, e; 1) = \frac{\Gamma(d, e)}{\Gamma(-c + d, s + c)}\,{}_3 \\
F_2(c, -a + e, -b + e; s + c, d; 1),
\end{aligned}
\tag{8.31}
$$

where $s = e + d - a - b - c$, is the parameter excess.

9. If we start with the transformation formula in Gasper and Rahman (1990, (3.4.8), with $q \to 1$, reversed)

$$
\begin{aligned}
&(1 + z)\ {}_2F_1\left(a, \frac{1}{2} + a; b; -\frac{4z}{(1 - z)^2}\right) \\
&= (1 - z)^{2a}{}_3F_2\left(2a - 1, \frac{1}{2}\right) + a, 2a - b; -\frac{1}{2}, b; -z,
\end{aligned}
\tag{8.32}
$$

and assume that a is a non-positive integer, then we obtain

$$(d+e){}_4F_3\left(a, \frac{1}{2}+a, d, -e; b, \frac{1}{2}+\frac{d}{2}-\frac{e}{2}, 1+\frac{d}{2}-\frac{e}{2}; 1\right)$$
$$= \frac{\Gamma(1+e, 2a-d-e)}{\Gamma(2a+e, e-d)} \tag{8.33}$$
$${}_4F_3\left(2a-1, \frac{1}{2}+a, 2a-b, d; -\frac{1}{2}+a, b, 2a+e; -1\right),$$

provided a or d is a non-positive integer. (In fact, when applying the beta integral method, we have to deal with a sum of two series on the left-hand side, which generates the factor $(d+e)$ in the result.)

10. If we start with the transformation formula (Bailey 1935, p 97, example 4 (iv), with $x \to z$)

$${}_3F_2\left(\frac{1}{2}a, \frac{1}{2}+\frac{a}{2}, 1+a-b-c; 1+a-b, 1+a-c; -\frac{4z}{(1-z)^2}\right) \tag{8.34}$$
$$=(1-z)^a\ {}_3F_2(a, b, c; 1+a-b, 1+a-c; z)$$

and assume that a is a non-positive integer, then we obtain

$${}_5F_4\left(\begin{matrix}\frac{1}{2}+\frac{a}{2}, \frac{a}{2}, 1+a-b-c, 1-e \\ 1+a-b, 1+a-c, \frac{1}{2}+\frac{d}{2}-\frac{e}{2}\end{matrix}; 1\right) \tag{8.35}$$
$$= \frac{\Gamma(e, a-d+e)}{\Gamma(a+e, e-d)}\ {}_4F_3(a, b, c, d; 1+a-b, 1+a-c, a+e; 1),$$

provided a or d is a non-positive integer. This is a known transformation between a nearly-poised ${}_4F_3(1)$ series and a Saalschützian ${}_5F_4(1)$ series (see Slater 1966, (2.4.2.3)—to see this do the replacements: $a \to f, b \to 1+f-h, c \to h-a, e \to g-f$, in (8.35)).

11. If we start with the transformation formula (Bailey 1935, p 97, example 6, with $b \to 1+a-b$, $c \to 1+a-c$, $x \to z$):

$$(1+z){}_3F_2\left(\frac{1}{2}+\frac{a}{2}, 1+\frac{a}{2}, -a+b+c-1; b, c; -\frac{4z}{(z-1)^2}\right) \tag{8.36}$$
$$=(1-z)^{1+a}{}_4F_3\left(a, 1+\frac{a}{2}, 1+a-b, 1+a-c; \frac{a}{2}, b, c; z\right)$$

and assume that a is a non-positive integer, then we obtain

$$
{}_5F_4\left(\begin{matrix} \frac{1}{2}+\frac{a}{2},\ 1+\frac{a}{2},\ -a+b+c-1,\ d,\ -e \\ b,\ c,\ \frac{1}{2}+\frac{d}{2}-\frac{e}{2},\ 1+\frac{d}{2}-\frac{e}{2} \end{matrix};1\right)
= \frac{1}{d+e}\frac{\Gamma(1+e,\ 1+a-d+e)}{\Gamma(1+a+e,\ e-d)}
\times {}_5F_4\left(\begin{matrix} a,\ 1+\frac{a}{2},\ 1+a-b,\ 1+a-c,\ d \\ \frac{a}{2},\ b,\ c,\ 1+a+e \end{matrix};1\right), \tag{8.37}
$$

provided a or d is a non-positive integer. On replacing $b \to 1+a-b$, $c \to 1+a-c$, $d \to -m$, $e \to 1+a-w$, this identity corresponds to Bailey (1935, (4.5.2)). (The parenthetical remark in item 9 applies also here. However, in the result, we moved the factor $(d+e)$, which is generated by the sum of two series' on the left-hand side of (3.21), to the right-hand side.)

12. If we start with the transformation formula of Andrews and Stanton (1998, (5.8))

$$
\left(1-\frac{x}{2}\right){}_4F_3\Bigg(1+a,\ a-2b,\ 1-a+2b,\ \frac{1}{3}(a+4);
\left(\frac{3}{2}+b,\ 1+a-b,\ \frac{1}{3}(a+1);\ -\frac{z^2}{4(1-z)}\right)
= (1-2z)(1-z)^{1+a}
\times {}_4F_3\left(1+a,\ 1+b,\ \frac{1}{2}+a-b,\ \frac{1}{3}(5+2a)\right);
2+2b,\ 1+2a-2b,\ \frac{2}{3}(1+a;\ 4(1-z)z), \tag{8.38}
$$

and assume that a is a negative integer, then we obtain

$$
{}_6F_5\left(\begin{matrix} \frac{1}{3}(4+a),\ 1+a,\ a-2b,\ 1-a+2b,\ \frac{1}{2}(1+d),\ \frac{d}{2} \\ \frac{1}{3}(1+a),\ 1+a-b,\ \frac{3}{2}+b,\ 1+d-e,\ 1+e \end{matrix};1\right)
= \frac{2(1+a-2d+e)}{2e-d}\frac{\Gamma(1+e,\ 1+a-d+e)}{\Gamma(2+a+e,\ e-d)}
\times {}_6F_5\left(\begin{matrix} \frac{1}{3}(5+2a),\ 1+a,\ \frac{1}{2}+a-b,\ 1+b,\ d,\ 1+a-d+e \\ \frac{2}{3}(1+a),\ 1+2a-2b,\ 2+2b,\ 1+\frac{1}{2}(a+e),\ \frac{3}{2}+\frac{1}{2}(a+e) \end{matrix};1\right), \tag{8.39}
$$

provided a is a negative integer or d is a non-positive integer. (Again, the parenthetical remark after item 9 applies, this time on both sides. The factors generated appear in the first term on the right-hand side.)

13. If we start with the cubic transformation formula of Bailey (1928, (4.05), with $\rho_1 \to b$, $\rho_2 \to 3a - b + \frac{3}{2}$, $x \to z$)

$$(1-z)^{-3a}{}_3F_2\left(a, \frac{1}{3}+a, \frac{2}{3}+a; b, \frac{3}{2}+3a-b; -\frac{27z}{4(1-z)^3}\right) = {}_3F_2\left(3a, -3a+2b-1, 2+3a-2b; b, \frac{3}{2}+a-b; \frac{z}{4}\right) \tag{8.40}$$

and assume that a is a non-positive integer, then we obtain

$$\begin{aligned} &{}_6F_5\left(\begin{matrix} a, \frac{1}{3}+a, \frac{2}{3}+a, d, \frac{1}{2}(1+3a-e), 1+3a-e \\ \frac{1}{2}(3+3a-b), b, \frac{1}{3}(1+3a+d-e), \frac{1}{3}(2+3a+d-e), \frac{1}{3} \\ (3+3a+d-e) \end{matrix}; 1\right) \\ &= \frac{\Gamma(-3a+e, e-d)}{\Gamma(e, -3a-d+e)} \\ &\times {}_4F_3\left(3a, 2+3a-2b, -3a+2b-1, d; \frac{3}{2}+3a-b, b, e; \frac{1}{4}\right), \end{aligned} \tag{8.41}$$

provided a or d is a non-positive integer. This is an unusual identity featuring a transformation between a ${}_6F_5(1)$ series and a ${}_4F_3(1)$ series.

14. If we start with the second cubic transformation formula of Bailey (1928, (4.06), with $\rho_1 \to b$, $\rho_2 \to 3a - b + \frac{3}{2}$, $x \to z$)

$$(1-z)^{-3a}{}_3F_2\left(a, \frac{1}{3}+a, \frac{2}{3}+a; b, \frac{3}{2}+3a-b; \frac{27z^2}{4(1-z)^3}\right) = {}_3F_2\left(3a, -\frac{1}{2}+b, 1+3a-b; 2b-1, 2+6a-2b; 4z\right) \tag{8.40}$$

and assume that a is a non-positive integer, then we obtain

$$\begin{aligned} &{}_6F_5\left(\begin{matrix} a, \frac{1}{3}+a, \frac{2}{3}+a, \frac{1}{2}(1+d), \frac{d}{2}, e \\ \frac{3}{2}(3a-b), b, \frac{1}{3}(d+e), \frac{1}{3}(1+d+e), \frac{2}{3}(1+d+e) \end{matrix}; 1\right) \\ &= \frac{\Gamma(1-e, 1+3a-d-e)}{\Gamma(1+3a-e, 1-d-e)} \\ &\times {}_4F_3\left(3a, 1+3a-b, -\frac{1}{2}+b, d; 2+6a-2b, 2b-1, 1+3a-e; 4\right), \end{aligned} \tag{8.41}$$

provided a or d is a non-positive integer.

15. If we start with the transformation formula (Berndt 1989), entry four of Ramanujan in his notebook (2, chapter 11, p 50 $\beta \to b$, $x \to z$) and Rahman and Verma (1993, (5.12), with $z \to z/(2-z)$)

$$
\begin{aligned}
&{}_2F_1\left(\frac{1}{2}a, \frac{1}{2}(1+a); \frac{1}{2}+b; -\frac{4z}{(1-z)^2}\right)\\
&=(1-z)^a {}_2F_1\left(a, \frac{1}{2}+a-b; \frac{1}{2}+b; -z\right)
\end{aligned}
\tag{8.42}
$$

and assume that a is a non-positive integer, then we obtain

$$
\begin{aligned}
&{}_4F_3\left(\frac{1}{2}(1+a), \frac{a}{2}, d, e; \frac{1}{2}+b, \frac{1}{2}(d+e), \frac{1}{2}(1+d+e); 1\right)\\
&=\frac{\Gamma(1-e, 1+a-d-e)}{\Gamma(1+a-e, 1-d-e)}\\
&\times {}_3F_2\left(a, \frac{1}{2}+a-b, d; \frac{1}{2}+b, 1+a-e; -\right),
\end{aligned}
\tag{8.43}
$$

provided a or d is a non-positive integer.

16. The transformation formula (Slater 1966, (1.8.10)), with $c \to 1+a-b-c$ and $z \to 1-z$, expressing the Gauss solution valid for $|z| < 1$ in terms of the Gauss functions valid for $|z-1| < 1$ is

$$
\begin{aligned}
{}_2F_1(a, b; c; z) = (1-z)^{c-a-b}&\frac{\Gamma(c, a+b-c)}{\Gamma(a, b)}\\
&\times {}_2F_1(-b+c, -a+c; 1-a-b+c; 1-z)\\
&+\frac{\Gamma(c, c-a-b)}{\Gamma(c-a, c-b)} {}_2F_1(a, b; 1+a+b-c; 1-z),
\end{aligned}
\tag{8.44}
$$

which, on applying the beta integral method algorithm, results in

$$
\begin{aligned}
{}_3F_2(a, b, c; d, e; 1) = &\frac{\Gamma(d, d-a-b)}{\Gamma(d-a, d-b)}\\
&{}_3F_2(a, b, -c+e; 1+a+b-d, e; 1)\\
&+\frac{\Gamma(a+b-d, d, e, d+e-a-b-c)}{\Gamma(a, b, e-c, d+e-a-b)}\\
&\times {}_3F_2(d-a, d-b, d+e-a-b-c;\\
&1+d-a-b, d+e-a-b; 1).
\end{aligned}
\tag{8.45}
$$

This three-term ${}_3F_2(1)$ formula can be found in Slater (1966, (4.3.4.2)).

17. The transformation formula given in Bruce Berndt (1989), as entry 21 of Ramanujan:

$$
\begin{aligned}
{}_2F_1\left(a, b; \frac{1}{2}(1+a+b); \frac{1-z}{2}\right) &= z\frac{\Gamma\left(-\frac{1}{2}(1+a+b)\right)}{\Gamma(a/2, b/2)} \\
&\times {}_2F_1\left(\frac{1}{2}(1+a), \frac{1}{2}(1+b); 3/2; z^2\right) \\
&+ \frac{\Gamma\left(\frac{1}{2}, \frac{1}{2}(1+a+b)\right)}{\Gamma\left(\frac{1}{2}(1+a), \frac{1}{2}(1+b)\right)} {}_2F_1(a/2, b/2; 1/2; z^2),
\end{aligned}
\tag{8.44}
$$

results in

$$
\begin{aligned}
&{}_3F_2\left(a, b, e-d; \frac{1}{2}(1+a+b), e; \frac{1}{2}\right) \\
&= \frac{\Gamma\left(\frac{1}{2}, \frac{1}{2}(1+a+b)\right)}{\Gamma\left(\frac{1}{2}(1+a), \frac{1}{2}(1+b)\right)} \\
&\times {}_4F_3\left(\frac{a}{2}, \frac{b}{2}, \frac{1}{2}(1+d), \frac{d}{2}; \frac{1}{2}, \frac{1}{2}(1+e), \frac{e}{2}; 1\right) \\
&+ \frac{\Gamma\left(-\frac{1}{2}, \frac{1}{2}(1+a+b), 1+d, e\right)}{\Gamma(a/2, b/2, d, 1+e)} \\
&\times {}_4F_3\left(\frac{1}{2}(1+a), \frac{1}{2}(1+b), \frac{1}{2}(1+d), 1+\frac{d}{2}; \right. \\
&\left. \frac{3}{2}, \frac{1}{2}(1+e), 1+\frac{e}{2}; 1\right).
\end{aligned}
\tag{8.45}
$$

To conclude, we have shown in this chapter how we can systematically generate new hypergeometric transformations from older transformations by a procedure called the beta integral method.

References

Andrews G E, Askey R and Roy R 1999 *Special Functions* (Cambridge: Cambridge University Press)

Andrews G E and Stanton 1998 Determinants in plane partition enumeration *Europe. J. Combin.* **19** 273

Bailey W N 1928 *Proc. London Math. Soc.* **2** 242

Bailey W N 1935 *Generalized Hypergeometric Series* (Cambridge: Cambridge University Press)

Berndt B 1989 *Ramanujan's Notebooks, Part III* (Berlin: Springer)

Erdélyi A 1939 Transformation of hypergeometric integrals by mean of fractional integration by parts *Quart. J. Math. Oxford* **10** 176–89

Gasper G 1975 Formulas of Dirchlet-Mehler type *Lecture Notes in Mathematics* **457** 207

Gasper G and Rahmann M 1991 *Basic Hypergeometric Series* (Cambridge: Cambridge University Press). Also, ed Gasper G and Rahmann M 1990 *Encyclopedia of Mathematics and its Applications* vol 35 (Cambridge: Cambridge University Press)

Gasper G 2000 q-Extensions of Erdélyi's fractional integral representations for hypergeometric functions and some summation formulas for double q-Kampé de Fériet series *Contemp. Math.* **254** 187

Joshi C M and Vyas V 2003 *J. Comput. Appl. Math.* **160** 125

Krattenthaler C and Srinivasa Rao K 2003 *J. Comput. Math. Appl.* **160** 159

Krattenthaler C 1995 *Symbol J. Comput.* **20** 737

Luke Y L 1969 *The Special Functions and their Applications* vol I (London: Academic)

Petkovsek M and Salvy B 1993 *Proc. ISSAC* **93** 27–33

Rahman M and Verma A 1993 *Trans. Amer. Math. Soc.* **335** 277

Singh V N 1959 The basic analogues of identities of the Cayley-Orr type *J. London Math. Soc.* **34** 15

Slater L J 1966 *Generalized Hypergeometric Functions* (Cambridge: Cambridge University Press)

Srivastava H M and Karlsson Per W 1985 *Multiple Gaussian Hypergeometric Series* (Chichester: Ellis Horwood), p 270

Weber M and Erdélyi A 1952 *Amer. Math. Monthly* **59** 163

Wilf H S and Zeilberger D 1992 *Invert. Math.* **108** 575

IOP Publishing

Generalized Hypergeometric Functions
Transformations and group theoretical aspects
K Srinivasa Rao and Vasudevan Lakshminarayanan

Chapter 9

Gauss, hypergeometric series, and Ramanujan

9.1 Introduction

As discussed in chapter 1, Carl Friedrich Gauss discovered the second order ordinary differential equation (1.29), characterized by three regular singular points at 0, 1, ∞, and the summation theorem satisfied by ${}_2F_1(a; b; c; 1)$, for arbitrary complex values of a, b, and c. He published his comprehensive and renowned thesis on the hypergeometric series, entitled *Disquisitiones Generales Circa Seriem Infinitam*. Gauss was prolific in mathematics and physics, and made innumerable contributions. His works—Gauss's *Werke*—run to some 23 volumes and are preserved in Göttingen for posterity. Generations of mathematicians have been researching and extending the research of Gauss to this day. Besides his *Werke,* Gauss was a great communicator and wrote about 6000 letters to friends and relatives and these are preserved and are the subject matter of innumerable studies. See Dunnington (1955) for a technical biography.

Gauss rose to prominence and lived for 78 years. During his lifetime he was duly recognized for his stupendous talents; he was the astronomer royal, in-charge of the mint in later years of his life. He was, on occasion, introspective and one of his famous statements about his childhood is that 'he could count before he could talk', and helped his father correct a mistake while paying wages to the workers of their business house for flax. His contemporaries considered him a titan.

To get a glimpse into the stature and nature of Gauss, we reproduce one incident in his life (Srinivasa Rao and Vanden Berghe 2004).

In 1816, when Olbers[1] drew the attention of Gauss to Fermat's last theorem that there are no integer solutions to the Diophantine equation

$$x^n + y^n = z^n, \qquad \text{for } n > 2, \tag{9.1}$$

[1] Heinrich Wilhelm Olbers (1758–1840), German astronomer. Known for what is called Olber's paradox or the 'dark night sky' paradox.

for which Fermat had a simple proof—Gauss replied, 'I am quite obliged for your report of the Paris prize. But I must say that Fermat's theorem considered as an isolated proposition, interests me very little; I could very easily propose a whole string of such propositions, which no one should be able to prove or use.' Sophie Germain[2] made a significant advance when she suggested a method to show that Fermat's last theorem is true when n is a prime p such that $2p + 1$ is also a prime (for instance, when $p = 5$, $2p + 1 = 11$, is also a prime). Germain communicated her results and had correspondence with Gauss, but she signed letters using a pseudonym, Monsieur Le Blanc. However, in 1806, Gauss came to know the true identity of the lady mathematician. The story of how her identity was revealed, by the invading General of the French Army, to whom Germain wrote asking for a guarantee to Gauss' safety, is narrated by Simon Singh (1998). Impressed by Germain's breakthrough and her anxiety for his safety at the hands of the army, Gauss warmly wrote in a letter to her:

> A taste for the abstract sciences in general and above all the mysteries of numbers is excessively rare: one is not astonished at it: the enchanting charms of this sublime science reveal themselves only to those who have the courage to go deeply into it. But when a person of the sex which, according to our customs and prejudices, must encounter infinitely more difficulties than men to familiarize herself with these thorny researches, succeeds nevertheless in surmounting these obstacles and penetrating the most obscure parts of them, then without doubt must have the noblest courage, quite extraordinary talents and superior genius.

Such chivalrous correspondence as this from Gauss encouraged and inspired the lady Germain. However, after becoming the director of the astronomy observatory, Gauss' interest in number theory waned and the correspondence did not continue. Still, years later, Gauss convinced the University of Göttingen to grant Germain an honorary degree. But tragically she died of breast cancer at the age of 55 before the honor was conferred on her. Regarding this episode, the Hungarian mathematician, Paul Erdös,[3] wrote: 'being nice to her [Germain] was not typical behavior of Gauss ... Gauss was mean, although not as mean as Newton.... Often when students shared their work with Gauss, he would tell them he had done it all before. Maybe he had, may be he hadn't. But it was wrong of him to squash the youthful enthusiasm of students.' As for Fermat's last theorem, it was only in 1995 that Andrew Wiles dramatically announced the solution of this 300 year old problem with Fermat (Wiles 1995).

Srinivasa Ramanujan

'Srinivasa Ramanujan was a mathematician so great that his name transcends jealousies, the one superlatively great mathematician whom India produced in the

[2] Mane-Sophie Germain (1776–1831), French mathematician. See D E Musielak (2015).
[3] Paul Erdos, Hungarian Mathematician (1913–1996). He was a periptatic and prolific mathematician (about 1525 papers in his lifetime with over 500 collaborators; see Hoffman 1998.

last thousand years.' Thus began a broadcast, in Hindustani, in 1941, by visiting professor to the University of Madras, Prof E H Neville, from Cambridge, who came to teach complex analysis. He was also tasked by Prof G H Hardy (1940), his colleague at Trinity College, Cambridge, to find out the facts about Ramanujan. Hardy received an 11-page letter from Ramanujan in January 1913, containing the mathematical statements about 120 theorems without proofs. These were selected nuggets of Ramanujan to showcase his ability as a mathematician. Later on, Hardy considered, Ramanujan was his discovery though he has stated that, like all other great men, he [Ramanujan] invented himself. Hardy recorded 'I still know more about Ramanujan than any one else, am still the first authority on this particular subject. . . . I owe more to him than to any one else in the world with one exception, and my own association with him is the one romantic incident in my life'.

As noted in the previous paragraph, the Hardy–Ramanujan collaboration started with the historic letter from Ramanujan to Hardy. This had four series formulae about which Hardy commented: 'the series formulae I found much more intriguing, and it becomes obvious that Ramanujan must possess much more general theorems'. We will in this chapter refer to the discoveries of Ramanujan and his work on the hyper- geometric series; where, with only the hint of the Gauss summation theorem for ${}_2F_1(1)$ (1.18), he stated the most general summation theorem for ${}_7F_6(1)$, and, more astonishingly, derives it in one chapter in his celebrated notebooks. According to prof Richard A Askey, his notebooks contain 3254 theorem referred to as entries in the notebooks of Ramanujan.

Ramanujan started noting down entries in three notebooks, perhaps from the time he finished his schooling—matriculation examination at the University of Madras—by the end of 1903. Except for chapter 1, in Notebook 1, which has the title 'Magic Squares', none of the chapters have headings. The chapters were numbered with Roman numerals and the entries numbered sequentially with Arabic numbers. Some of the entries had several examples that were also numbered with lower case Arabic numbers.

It is a pity that the original notebooks have not been preserved for posterity, like the works of Newton in the Wren Library of Trinity College, Cambridge, UK.

At the time of the birth centenary of Ramanujan in 1987, one of us (KSR) located the notebooks in the safe custody of the main library at the University of Madras. Permission was sought and obtained from the vice chancellor and the registrar of the university to make photo-graphic images of the three notebooks. These are available on the website http://www.imsc.res.in/~rao/ramanujan.

Thanks to the concerted efforts of Dr Ranganathan, who went along with Dr Homi J Bhabha and Dr K S Krishnan to the then Prime Minister of India, Pundit Jawaharlal Nehru—a limited facsimile edition (2000 copies) of the notebooks of Ramanujan, was produced in 1957 by the Tata institute of fundamental research, Mumbai, India, to celebrate the 70th anniversary of Ramanujan.

However, as for the originals, the pages in the leather bound notebooks have become brittle, we urge the university authorities to preserve these national mathematical treasures of the country.

The notebooks of Ramanujan have been reprinted again, in an excellent technical production, and sponsored by the National Board for Higher Mathematics (India) on the occasion of the 125th Birthday anniversary of Ramanujan, in 2012. On that day, the prime minister of India declared that, from then on, 22nd December will be National Mathematics Day in India.

9.2 On some entries of Ramanujan on hypergoemtric series in his notebooks

G H Hardy *et al* (1927) state in a footnote on p xxv in his *Introduction to the Collected Papers of Srinivasa Ramanujan* (1927), that 'there is always more in one of Ramanujan's formulas than meets the eye, as anyone who sets to work to verify those which look the easiest will soon discover. In some the interest lies very deep, in others comparatively near the surface; but there is not one which is not curious or entertaining.'

Ramanujan's first four results presented to Hardy, in his letter of January 16, 1913, were the following:

$$1-\frac{3!}{(1!2!)^3}x^2+\frac{6!}{(2!4!)^3}-\cdots=\left(1+\frac{x}{(1!)^3}+\frac{x^2}{(2!)^3}\right)\times\left(1-\frac{x}{(1!)^3}+\frac{x^2}{(2!)^3}-\cdots\right), \tag{9.2}$$

$$1-5\left(\frac{1}{2}\right)^3+9\left(\frac{1\cdot 3}{2\cdot 4}\right)^3-13\left(\frac{1\cdot 3\cdot 5}{2\cdot 4\cdot 6}\right)^3+\cdots=\frac{2}{\pi}, \tag{9.3}$$

$$1+9\left(\frac{1}{4}\right)^4+17\left(\frac{1\cdot 5}{4\cdot 8}\right)^4+25\left(\frac{1\cdot 5\cdot 9}{4\cdot 8\cdot 12}\right)^3+\cdots=\frac{2^{3/2}}{\pi^{1/2}\{\Gamma(3/4)\}^2}, \tag{9.4}$$

$$1-5\left(\frac{1}{2}\right)^5+9\left(\frac{1\cdot 3}{2\cdot 4}\right)^5-13\left(\frac{1\cdot 3\cdot 5}{2\cdot 4\cdot 6}\right)^5+\cdots=\frac{2}{\Gamma(3/4)^4}. \tag{9.5}$$

Hardy (1927) stated that 'the series formulae I found much more intriguing, and it becomes obvious that Ramanujan must possess much more general theorems. The second is a formula of Bauer well known in the theory of the Legendre series, but the others are much harder than they look. The theorems required for proving them can all be found now in Bailey's Cambridge tract (1935) on hypergeometric functions'.

Ramanujan had his own method of discovery for deep mathematical results without a formal education and with often only a hint. Hardy, his mentor, who invited Ramanujan to Trinity College, Cambridge, for a five year stay (in England from 14 April 1914 to 27 February 1919), said that whenever he wanted to convey a new result Ramanujan, would point out several more. Also, when Hardy asked for proof of any entry in the notebooks of Ramanujan, he would be provided with four or five different proofs of the entry. His methods were simpler and novel. To give an example of this aspect of Ramanujan, we provide his proof of the Gauss summation theorem (1.19), which was proved in chapter 1 using the Euler beta integral representation for ${}_2F_1$ (1.18).

Ramanujan's entry 8 in chapter X of Notebook 2 is:

If $\Re(x + y + n + 1) > 0$, then

$$ {}_2F_1(-x, -y; n + 1; 1) = \frac{\Gamma(n + 1)\Gamma(x + y + n + 1)}{\Gamma(x + n + 1)\Gamma(y + n + 1)}. \tag{9.6} $$

Proof: assume n, x are integers $n \geqslant 0$, $x \geqslant 0$, and let

$$ (1 + u)^{y+n} = \sum_k {}^{y+n}C_k\, u^k = \quad \text{and} \left(1 + \frac{1}{u}\right)^x = \sum_\ell {}^xC_\ell \left(\frac{1}{u}\right)^\ell, $$

$$ (i) \quad (1 + u)^{y+n}\left(\frac{1}{u}\right)^x = \sum_{k,\ell} {}^{y+n}C_k\; {}^xC_+\ell\; u^{k-\ell} = \sum_n a_n\, u^n, \text{ (say)}, $$

then

$$ a_n = \sum_\ell {}^{y+n}C_{\ell+n}\,{}^xC_\ell = \frac{\Gamma(y + n + 1)}{\Gamma(n + 1)\Gamma(y + 1)}\,{}_2F_1(-x, -y; n + 1; 1) $$

$$ (i) \quad \frac{(1 + u)^{x+y+n}}{u^x} = \sum_\ell {}^{x+y+n}C_{\ell+n}\, u^{\ell-x} = \sum_n a_n\, u^n, \text{ (say)}, $$

then

$$ a_n = {}^{x+y+n}C_{x+n} = \frac{\Gamma(x + y + n + 1)}{\Gamma(x + n + 1)\Gamma(y + 1)}. $$

Equating the coefficients,

$$ \frac{\Gamma(y + n + 1)}{\Gamma(x + n + 1)\Gamma(y + 1)}\,{}_2F_1(-x, -y; n + 1; 1) = \frac{\Gamma(x + y + n + 1)}{\Gamma(x + n + 1)\Gamma(y + 1)} $$

and hence,

$$ {}_2F_1(-x, -y; n + 1; 1) = \frac{\Gamma(n + 1)\,\Gamma(x + y + n + 1)}{\Gamma(x + n + 1)\,\Gamma(y + n + 1)}, $$

which, for $x \to -a$, $y \to -b$, $n \to c - 1$, is the Gauss summation theorem:

$$ {}_2F_1(a, b; c; 1) = \frac{\Gamma(c)\Gamma(c - a - b)}{\Gamma(c - a)\Gamma(c - b)}. \tag{1.18} $$

Ramanujan discovered for himself all the classical summation theorems of Gauss, Chu-Vandermonde, Kummer, Pfaff–Saalschütz, Dixon, and Dougall. In fact, Dougall's ${}_7F_6$ summation theorem is entry 1 in chapter XII of Notebook 1, and also entry 1 in chapter X of Notebook 2. This example underscores the fact that the second notebook is not a copy of the first notebook of Ramanujan, but it is considered a revised, enlarged version of the first notebook. This ${}_7F_6$ summation theorem, discovered by Dougall (1907), was independently discovered by Ramanujan, perhaps during the 1910–12 period. There are no dates anywhere in the notebooks, so it is not possible to assert the dates of discovery of any of the 3254 entries/theorems of Ramanujan contained in his notebooks. In any case, the experts consider Ramanujan to have rediscovered not only all that was known in Europe, but discovered several new theorems. In particular, theorems on products of the hypergeometric series (Slater 1966) as well as several types of asymptotic expansions.

On the other hand it is also well known (Berndt 1987, chapter X, pp 7–47) that Ramanujan did not publish any of his results from his chapters on the hypergeometric series in his notebooks, even though he had them with him when he was in Cambridge. Chapters X and XI in the Notebook 2 have been extensively studied by Hardy (1940) and Berndt (1987, ch. X).

Our aim is to provide a proof for example 7, after entry 43, in chapter XII of Notebook 1—which does not find a place in the corresponding examples after entry 10 of Notebook 2, and perhaps that is why it is not discussed by Berndt (1987).

9.3 Entry 43, in chapter XII of Ramanujan's Notebook 1

As stated above chapter XII of Notebook 1 and chapter X of Notebook 2, both notebooks are on the hypergoemetric series, and they start with the most general theorem known to date: the ${}_7F_6$ theorem. This has since been christened the Dougall–Ramanujan ${}_7F_6$ summation theorem by G H Hardy, who said that the order of the names is due to the fact that while Dougall's paper was published in 1909, Ramanujan's entries in his notebooks are not dated and perhaps noted down by Ramanujan during 1904–13.

Of the 3254 entries and theorems, only about 5% belong to the realm of elementary mathematics, and chapters XII of Notebook 1 and X of Notebook 2, are on the hypergoemetric series. What is remarkable is that, in both chapters, he discovers all the known summation theorems of the hypergeometric series with only a hint of the Gauss ${}_2F_1$ summation theorem (1812), and starts the chapters with the most general summation theorem, known as the 1909 Dougall summation theorem referred above.

The entry 43 is

$$\begin{aligned}&\frac{\pi}{\tan(\pi x)}\frac{\|\underline{2x}}{(2^x\|\underline{x})^2\,(1-2x)}\left(\sum\frac{1}{2x}-\frac{1}{2}\sum\frac{1}{x}+\frac{1}{1-2x}-\frac{\pi}{2}\tan(\pi x)\right)\\&=\frac{1}{1^2}+\frac{x}{|\underline{1}}\cdot\frac{1}{3^2}+\frac{x(x+1)}{|\underline{2}}\cdot\frac{1}{5^2}+\&c.\qquad (\mathit{NB}1,\ \mathit{ch.}\ \mathit{XII},\ \mathit{entry}\ 43,\ \mathit{ex}.7).\end{aligned}\tag{9.6}$$

It is to be noted that Ramanujan did not include example 7 after entry 43 in chapter XII of his first notebook in his second notebook, while all the other examples after entry 43 found their way into the second notebook. Srinivasa Rao *et al* (2006) have derived an equivalent theorem, and, in fact, an identity that is more general than example 5 (of section 10 of chapter 10 of Notebook 2). It is to be noted that the above equation is in the notation of Ramanujan, where $\lfloor u(x)$ is for the gamma function $\Gamma(x+1) = (x+1)!$, a function over real numbers.

The factor on the left-hand side of (9.6)

$$\frac{\pi}{\tan(\pi x)} \frac{\lfloor 2x}{(2x \lfloor x)^2 (1-2x)} \tag{9.7}$$

is the same that appears also on the left side of example 4, following entry 43 in chapter XII of Notebook 1:

$$\frac{\pi}{\tan(\pi x)} \frac{\lfloor 2x}{(2x \lfloor x)^2 (1-2x)} = 1 + \frac{x}{\lfloor 1} \frac{1}{3} + \frac{x(x=1)}{\lfloor 2} \frac{1}{5} + \&c. \tag{9.8}$$

(*NB*1, *Ch. XII*, *entry* 43, *ex*.4)

This is given in Notebook 2 as

$$\frac{\pi \lfloor n}{2 \lfloor n + \frac{1}{2}} = 1 - \frac{n}{\lfloor 1} \cdot \frac{1}{3} + \frac{n(n-1)}{\lfloor 2} \cdot \frac{1}{5} + \&c. \tag{9.9}$$

(*NB*2, *ch. X*, *entry* 43, *ex*.4)

Factor (9.7) can be shown to be equal to

$$\frac{\sqrt{\pi}\Gamma(1-x)}{2x\,\Gamma(\frac{3}{2}-x)} = \frac{\pi\sqrt{\lfloor -x}}{2x\lfloor -x+1/2}, \tag{9.10}$$

after using the reflection formula

$$\Gamma(z)\Gamma(1-z) = \frac{\pi}{\sin(\pi z)} = \pi \operatorname{cosec}(\pi z) \tag{9.11}$$

and the duplication formula

$$\Gamma(2z) = 2^{2z-1}\pi^{-1/2}\Gamma(z)\Gamma z + \frac{1}{2} \tag{9.12}$$

and some algebraic simplifications.

When written in the standard hypergeometric notation, which we recall here as

$${}_rF_s\left[\begin{matrix} a_1, a_2, \dots, a_r \\ b_1, b_2, \dots, b_s \end{matrix}; z\right] = \sum_{k=0}^{\infty} \frac{(a_1)_k (a_2)_k \cdots a_r(k)}{(b_1)_k (b_2)_k \cdots, (b_s)_k} \frac{z^k}{k!} \tag{9.13}$$

where $(\alpha)_k$ is the Pochammer symbol, the series on the right-hand side is

$$
{}_2F_1\left(\frac{1}{2}, x; \frac{3}{2}; 1\right) = \frac{\Gamma\left(\frac{3}{2}\right)\Gamma\left(\frac{3}{2} - \frac{1}{2} - x\right)}{\Gamma\left(\frac{3}{2} - x\right)} = \frac{\sqrt{\pi}\Gamma(1 - x)}{2\Gamma\left(\frac{3}{2} - x\right)}
= \frac{\sqrt{\pi}}{2}\frac{\lfloor -x}{\lfloor -x + 1/2}, \tag{9.14}
$$

by the Gauss summation theorem (1.18). A comparison of these results with Ramanujan's entry shows that Ramanujan missed a multiplicative factor x on the left-hand side of his entry 43, example 4, in chapter XII of Notebook 1, while in his entry in chapter 10 of Notebook 2, it is is correct.

The series on the right-hand side of (9.6) in hypergeometric function notation is

$$
{}_3F_2(x, 1/2, 1/2; 3/2, 3/2; 1). \tag{9.15}
$$

There are two factors on the left-hand side of (9.6). Besides factor (9.7), the other factor is

$$
\sum \frac{1}{2x} - \frac{1}{2}\sum \frac{1}{x} + \frac{1}{1 - 2x} - \frac{\pi}{2}\tan(\pi x). \tag{9.15}
$$

Ramanujan used the notation $\sum \frac{1}{x}$ to indicate the extension of the function

$$
1 + \frac{1}{2} + \frac{1}{3} + \cdots + \frac{1}{n} \tag{9.16}
$$

representing the harmonic numbers from positive integers n to real X. In other words, $\sum \frac{1}{x}$ is Ramanujan's notation for the *digamma function*

$$
\psi(x) := \frac{\Gamma'(x)}{\Gamma(x)}, \tag{9.17}
$$

the logarithmic derivative of the famma function, or, more precisely

$$
\sum \frac{1}{x} = \psi(x + 1) + \gamma, \tag{9.18}
$$

where γ is the Euler–Mascheroni constant,[4] that is

$$
\psi(x + 1) = \sum \frac{1}{x} - \gamma. \tag{9.19}
$$

In fact, in Ramanujan's very first published research paper (Ramanujan 1911), the digamma function occurs as

[4] The Euler–Mascheroni constant (also called Euler's constant) occurs in number theory and analysis involving logarithmic and exponential integrals and has a value of 0.57721 56649.

$$\frac{d}{dn}\log\Gamma(n+1) = 1 + \frac{1}{2} + \frac{1}{3} + \cdots \frac{1}{n} - \gamma = \sum\frac{1}{n} - \gamma. \tag{9.20}$$

Thus, expression (9.15) becomes

$$\psi(2x+1) - \frac{1}{2}\psi(x+1) + \frac{1}{1-2x} - \frac{\pi}{2}\tan(\pi x) + \frac{\gamma}{2}. \tag{9.21}$$

The digamma function satisfies the recurrence relation (Erdélyi *et al* 1953, 1.7.1(8)):

$$\psi(z+1) = \psi(z) + \frac{1}{z}, \tag{9.22}$$

the reflection formula (Erdélyi *et al* 1953, 1.7.1(11)):

$$\psi(-z) = \psi(z+1) + \pi\cot(\pi z) \tag{9.23}$$

and the duplication formula (Erdélyi *et al* 1953):

$$2\psi(2x) = \psi(z) + \psi\left(z + \frac{1}{2}\right) + 2\log 2. \tag{9.24}$$

We remark that the duplication formula appears implicitly in Ramanaujan's first notebook. Namely, a comparison of entry 43, example 6 in Notebook 2, with $n \to x - 1$ in the entry (Notebook 1, chapter 10, example 6), which shows that Ramanujan obtained:

$$\sum\frac{1}{x+\frac{1}{2}} - \sum\frac{1}{x} = 2\sum\frac{1}{2x} + 2\sum\frac{1}{x} + \frac{1}{x} - 2\log 2, \tag{9.25}$$

which by (9.19) and (9.22) is equivalent to (9.24). Furthermore, the digamma function has the special values

$$\psi\left(\frac{1}{2}\right) = -\gamma - 2\log 2 \qquad \text{and} \qquad \psi(1) = -\gamma. \tag{9.26}$$

We now use the duplication formula (9.12) to convert (9.25) into

$$\frac{1}{2}\psi\left(x + \frac{1}{2}\right) + \log 2 + \frac{1}{1-2x} - \frac{\pi}{2}\tan(\pi x) + \frac{\gamma}{2}. \tag{9.27}$$

The recurrence relation (9.22) implies

$$\psi\left(x + \frac{1}{2}\right) + \psi\left(x - \frac{1}{2}\right) + \frac{2}{2x-1},$$

while we know from (9.26) that

$$\log 2 = -\frac{1}{2}\psi\left(\frac{1}{2}\right) - \frac{1}{2}\gamma. \tag{9.28}$$

If this is substituted in the last equation, and if we then apply the reflection formula (9.23), we obtain

$$\frac{1}{2}\left(\psi\left(\frac{3}{2}-x\right)-\psi\left(\frac{1}{2}\right)\right) \tag{9.29}$$

for the expression (9.15). Therefore, if we recall that factor (9.7) on the left-hand side of Ramanujan's entry (9.6)—in Notebook 1, chapter XII, entry 43, example 7—must be replaced by (9.14), the entry can be rewritten in the notation of Barnes (the contemporary notation for hypergeometric functions) as:

$${}_3F_2\left(x, \frac{1}{2}, \frac{1}{2}; \frac{}{3}2, \frac{3}{2}; 1\right) = \frac{\sqrt{\pi}}{4}\frac{\Gamma(1-x)}{\Gamma\left(\frac{3}{2}-x\right)}\left\{\psi\left(\frac{3}{2}-x\right)-\psi\left(\frac{1}{2}\right)\right\} \tag{9.30a}$$

$$\frac{1}{2}\left\{\psi\left(\frac{3}{2}-x\right)-\psi\left(\frac{1}{2}\right)\right\}{}_2F_1\left(x, \frac{1}{2}; \frac{3}{2}; 1\right), \tag{9.30b}$$

for $\Re x < 2$. In the view of Andrews, Askey, and Roy (1999, (1.2.13)),

$$\psi(x) - \psi(y) = \sum_{k=0}^{\infty}\left(\frac{1}{k+y} - \frac{1}{k+x}\right), \tag{9.31}$$

the equation (9.30*a*) is completely equivalent with Berndt (1987, part IV, second displayed line on p 410).

In the following section we state and prove the theorem that is a generalization of (9.30); that is, of Ramanujan's entry 43, example 7, of Notebook 1.

9.4 The theorem of Rao, Berghe, and Krattenthaler

Theorem. Let N be a non-negative integer and a a complex number that is not a negative integer. If $\Re x < N + 2$, then

$$\begin{aligned}{}_3F_2(a, a, x; a+1, a+N+1; 1) &= \frac{a\Gamma(a+N+1)\Gamma(1-x)}{n!\Gamma(a-x+1)} \\ &\times(\psi(a-x+1) - \psi(a) - \psi(N+1) - \gamma) \\ &- \frac{a\,\Gamma(a+N+1)\,\Gamma(1-x)}{n!\,\Gamma(a=x+1)}\sum_{k=1}^{N}\frac{(a)_k\,(-N)_k}{k\cdot k!\,(a-x+1)_k}.\end{aligned} \tag{9.32}$$

Proof. To evaluate the ${}_3F_2(1)$ series on the left-hand side of (9.32), we introduce a parameter ε, and we consider the series

$${}_3F_2(a, a-\varepsilon, x; 1+a-\varepsilon, 1+a-\varepsilon+N; 1).$$

We first apply the ${}_3F_2(1)$ non-terminating transformation formula (see Gasper and Rahmann 1990, ex. 3.6, $q \to 1$, reversed):

$$
\begin{aligned}
&{}_3F_2(A, B, C; D, E; 1) \\
&= \frac{\Gamma(A-B, D, E, s)}{\Gamma(A, D-B, E-B, s+B)} {}_3F_2(B, D-A, E-A; 1-A+B, s+b; 1) \\
&+ \frac{\Gamma(B-A, D, E, s)}{\Gamma(B, D-A, E-A, s+B)} {}_3F_2(A, D-B, E-B; s+A, 1+A-B; 1)
\end{aligned}
\tag{9.33}
$$

and obtain

$$
\begin{aligned}
&\frac{\Gamma(1+a-\varepsilon, \varepsilon, 1+a-\varepsilon+N, 2-\varepsilon+N-x)}{\Gamma(a, 1+N, 2+a-2\varepsilon, +N-x)} \\
&\times {}_2F_1(a-\varepsilon, 1-\varepsilon+N; 2+a-2\varepsilon+N-x; 1) \\
&+ \frac{\Gamma(1+a-\varepsilon, -\varepsilon, 1+a-\varepsilon+N, 2-\varepsilon+N-x)}{\Gamma(1-\varepsilon, a-\varepsilon, 1-\varepsilon+N, 2+a-\varepsilon+N-x)} \\
&\times {}_3F_2(a, 1, 1+N; 1+\varepsilon, 2+a-\varepsilon+N-x; 1)
\end{aligned}
\tag{9.34}
$$

where we have used the simplifying notation for the product of Gamma functions $\Gamma(x, y, \ldots) = \Gamma(x)\Gamma(y) \cdots$, and $s = D + E - A - B - C$.

Clearly, the convergence of the hypergeometric series on the right-hand side will only be guaranteed if $\Re x < 1$.

To the ${}_3F_2$-series, we then apply the transformation (see Bailey 1935, ex. 7, p 98), which is (4.102) belonging to the set of 10 non-terminating transformations presented in chapter 4:

$$
{}_3F_2\left(\begin{matrix} A, B, C \\ D, E \end{matrix}; 1\right) = \frac{\Gamma(E, s)}{\Gamma(s+A, E-A)} {}_3F_2\left(\begin{matrix} A, D-B, D-C \\ D, s+A \end{matrix}; 1\right),
\tag{9.35}
$$

to get the expression

$$
\begin{aligned}
&\frac{\Gamma(1+a-\varepsilon, \varepsilon, 1+a-\varepsilon+N, 2-\varepsilon+N-x)}{\Gamma(1, a, 1+N, 2+a-2\varepsilon+N-x)} {}_2F_1\left(\begin{matrix} a-\varepsilon, 1-\varepsilon+N \\ 2+a-2\varepsilon+N-x \end{matrix}; 1\right) \\
&+ \frac{\Gamma(1+a-\varepsilon, -\varepsilon, 1+a-\varepsilon+N, 1-x)}{\Gamma(1-\varepsilon, a-\varepsilon, 1-\varepsilon+N, +a-x)} {}_3F_2\left(\begin{matrix} a, \varepsilon, \varepsilon-N \\ 1+\varepsilon, 1+a-x \end{matrix}; 1\right).
\end{aligned}
\tag{9.36}
$$

The ${}_2F_1$-series is summed by means of the Gauss summation theorem (1.18), while the ${}_3F_2$-series is written as a sum over k, and subsequently split into the ranges $k = 0, k = 1, 2, \ldots, N$, and $k = N+1, N+2, \ldots$. This yields the expression

$$\begin{aligned}
&\frac{1}{\varepsilon}\left(\frac{\Gamma(1+a-\varepsilon, 1+\varepsilon, a+a-\varepsilon+N, 1-x)}{\Gamma(a, 1+N, 1+a-\varepsilon-x)}\right.\\
&\left.\frac{\Gamma(1+a-\varepsilon, 1+a-\varepsilon+N, 1-x)}{\Gamma(a-\varepsilon, 1-\varepsilon+N, 1+a-x)}\right)\\
&-\frac{\Gamma(1+a-\varepsilon, 1+a-\varepsilon+N, 1-x)}{\Gamma(a-\varepsilon, 1-\varepsilon+N, 1+a-x)_k}\sum_{k=1}^{N}\frac{(a)_k(\varepsilon-N)_k}{k!(k+\varepsilon)(1+a-x)_k}\\
&-\frac{\Gamma(1+a-\varepsilon, 1+a-\varepsilon+N, 1-x)}{\Gamma(a-\varepsilon, 1-\varepsilon+N, 1+a-x)}\sum_{k=N+1}^{\infty}\frac{(a)_k(\varepsilon-N)_k}{k!\,(k+\varepsilon)\,(1+a-x)_k}.
\end{aligned} \tag{9.37}$$

Now we take the limit $\varepsilon \to 0$. Thus, the original ${}_3F_2$-series becomes

$${}_3F_2(a, a, x; 1+a, 1+a+N; 1). \tag{9.38}$$

On the other hand, the four terms obtained simplify significantly. The last term simply vanishes because of the occurrence of the factor $(\varepsilon - N)_k$, which is equal to

$$(\varepsilon - N)_N\ \varepsilon(1+\varepsilon)_{k+N-1}, \text{ for } k \geqslant N+1 \to 0, \text{ as } \varepsilon \to 0, \tag{9.39}$$

for N, a non-negative integer.

9.5 Observations and concluding remarks

- The original notebooks of Ramanujan can be assessed at http://www.imsc.rao.in/~rao/ramanujan. See the original entry 43, example 7, on p 93 of Notebook 2.
- Theorem (9.32) reduces to Ramanujan's entry 43, example 7, in Notebook 2, in our notation (9.30), if $a = 1/2$ and $N = 0$.
- For $x = 1$, in (9.30), the ${}_3F_2(1)$ is a special case of Dixon's theorem in Slater (1966, III.8, for $a = 1$, $b = c = 1/2$), and it has the value $\pi^2/8$.
- For $x = 3/2$, the left-hand side of (9.30) is:

  $${}_3F_2(3/2, 1/2, 1/2; 3/2, 3/2; 1) = {}_2F_1(1/2, 1/2; 3/2; 1)\,\frac{\pi}{2}, \tag{9.39}$$

  by the Gauss summation theorem (1.18), which is the result of the right-hand side evaluated by l'Hôpital's rule.
- For $x = -k$, a negative integer, in (9.30), we get

  $${}_3F_2(1/2, 1/2, -k; 3/2, 3/2; 1) = \frac{\sqrt{\pi}}{2}\frac{\Gamma(k+1)}{\Gamma(k+3/2)}\sum_{j=1}^{k+1}\frac{1}{2j-1}. \tag{9.40}$$

  In view of (9.31), this is equivalent with example 5 of section 10, chapter 10 of Notebook 2 (Berndt 1987, part II, p 26).

- The ${}_3F_2(a, a, x; 1 + a, 1 + a + N; 1)$ series can be related to the 3-j coefficient

$$\begin{pmatrix} -\frac{x}{2} & -\frac{x}{2} & 0 \\ a - \frac{x}{2} & -a + \frac{x}{2} & 0 \end{pmatrix}$$

  (Srinivasa Rao 1978) provided x is a negative integer $-x \leqslant s \leqslant 0$. It also corresponds to the dual Hahn polynommial (Andrews *et al* 1999) $S_n(0; a, b = 1, c = 1 + N)$, for $x = -n$.

It should be noted that there is a missed multiplicative factor x on the left-hand side of entry 43, example 7, in chapter XII of Notebook 1. It is one of the very few instances in which such a trivial mistake occurs.

Professor Bruce Berndt states that though there are a few scattered errors in the notebooks, Ramanujan's accuracy is amazing and that mystery still surrounds some of his work, a statement which sums up the mathematical genius of Ramanujan.

These original notebooks have a history of their movement back and forth from India: Ramanujan took them with him to Cambridge in 1914 and brought them back in 1919 when he returned with them to Madras, with a BA (Cambridge), awarded for his 64-page long paper on highly composite numbers, submitted—along with half a dozen other published papers of his—as a thesis. He was also elected to the Fellowship of the Royal Society (FRS) and has the unique honour of being the first Indian mathematician to be so honoured. But after about a year-long stay in India, ailing with suspected tuberculosis which was a contagious dreaded disease in those days, he died on April 26, 1920. It was only in 1994, thanks to the research of the medical records with the TB Sanatoria in England, we now know that Ramanujan died of undiagnosed and so untreated hepatic amoebiasis, a tropical disease contacted by Ramanujan in 1906 in Madras (Young 1994). After his untimely death at the age of 32 years 4 months and 4 days, the Notebooks were sent back to professor Hardy. In March 1925, Hardy gave the notebooks of Ramanujan to Dr S R Rangnathan, a PhD (Cambridge) in library science, who was returning to India to become the librarian of the University of Madras, with the remark that 'Ramanujan belongs to your country, the proper place for his notebooks is your own (Madras) university library' (Srinivasa Rao 2004).

9.6 Epilogue

'The work of Ramanujan will be appreciated as long as people do mathematics', opined the astrophysicist Nobel Laureate Dr S Chandrasekhar at the time of the birth centenary of Ramanujan.

Undoubtedly, Srinivasa Ramanujan (December 22, 1887–April 26, 1920) is one of the greatest mathematicians of the twentieth century. For his mathematical abilities and natural genius he has been compared by his contemporaries, Professors

G H Hardy and J E Littlewood, with all-time great mathematicians, such as Leonhard Euler, Carl Friedrich Gauss, and Karl Gustav Jacobi.

In the spring of 1976, while going through the estate of the late Prof G N Watson, in the archives of Trinity College, George E Andrews (2012) discovered about 100 loose sheets of paper containing about 600 theorems, which he revealed as the lost notebook of Ramanujan (though it was neither lost nor was it a notebook). These theorems mostly dealt with what Ramanujan christened as mock theta functions. Andrews wrote a series of articles about the contents of these entries of Ramanujan during the last year of his life. It should be remarked that he was precariously ill but he still kept asking for paper from his young wife (who survived him and lived for another 74 years, after he died on April 26, 1920). George Andrews and Bruce Berndt published in three parts (2005–2013) their work, *Ramanujan's Lost Notebook and Other Unpublished Papers.* Ramanujan's 'Number theory work and classical analysis are in the spot light' is in the fourth of five projected volumes.

Any mention of Ramanujan's work cannot go without the oft repeated snippet about the taxi cab number 1729, by which Hardy had reached Ramanujan, confined at a TB Sanatorium in Putney—for nearly half of his five year stay in Cambridge, Ramanujan was in various sanatoria. Hardy's comment, that it was a dull number to start his conversation, was refuted by Ramanujan who said 'it is the smallest number that can be expressed as the sum of two cubes in two different ways', viz 123 + 13 = 103 + 93 (see, for instance, Srinivasa Rao 2004). This observation of Ramanujan made Littlewood comment that 'every positive integer was one of Ramanujan's personal friends'. The ability to determine patterns combined with good memory probably helped Ramanujan to come to important conclusions from a long complicated formula

The number theorist Ken Ono (2016), at the age of 16, brought tears to the eyes of his father, an admirer of Ramanujan, by asking him why he is being taken to task for dropping out of school (as Ramanujan did). In his book, Ono talks about Ramanujan's research work as a mathematical legacy to the world of mathematics; how a Fields Medal was awarded to Deligne in 1974 for solving the τ-conjecture of Ramanujan; that the taxi cab number appears in connection with the Diophantine equation: $a^3 + b^3 = c^3 \pm 1$; and the list goes on with mention of Ramanujan's pioneering work on partitions; Ramanujan graphs; the circle method and problems in number theory.

Ramanujan is an enigma[5] and, as Robert Kanigel (1991) in his biography of Ramanujan aptly refers to him, a Swayambhu (meaning self-manifested gift of God), inexplicable, just like the big bang origin of the universe. It was hard to understand the creative genius of Ramanujan who, without formal/proper education in mathematics, contributed so much to the field. Ramanujan, unquestionably, had enormous calculational powers; and we can consider him a great human calculator who had an uncanny ability to know if a particular mathematical relation was true.

[5] Nandan Khudyadi, an Indian filmmaker, produced the documentary, *The Enigma of Srinivasa Ramanujan*, on the birth centenary of Ramanujan followed by two sequels, *The Genius of Srinivasa Ramanujan* and *The Legacy of Srinivasa Ramanujan.* See Nandan Khudyadi productions.

More importantly, it can be argued that he had a fantastic (or even mysterious) ability to tell what was significant and what could not be deduced from it. We conclude this chapter with a quote from the probability theorist, Kac (1985):

> an ordinary genius is a fellow that you and I would be just as good as, if we were only many times better. There is no mystery as to how his mind works. Once we understand what he has done, we feel certain that we too, could have done it. It is different with the magicians. ... the working of their minds is for all intents and purposes incomprehensible. Even after we understand what they have done, the process by which they have done it is completely dark.

His quote refered to the great American physicist, Richard Feynmann (1918–1988; Feynman received the Nobel prize for his contributions to quantum electrodynamics and also developed a method which is now known as Feynman diagrams. he was also a great lecturer and educator; See Gleick (1993)), but is equally appropriate here.

References

Andrews G E 2012 The discovery of Ramanujan's lost notebook *The Legacy of Srinivasa Ramanujan: Proceedings of an International Conference in Celebration of the 125th Anniversary of Ramanujan's Birth* Univ. of Delhi. Also, 2012 The Discovery of Ramanujan's Lost Notebookfaculty.math.illinois.edu/~berndt/andrews.pdf

Andrews G E, Askey R and Roy R 1999 *Special Functions* (Cambridge: Cambridge University Press)

Bailey W N 1935 *Generalized Hypergeometric Series* Cambridge Tracts in Mathematics and Mathematical Physics 32 (Cambridge: Cambridge University Press)

Berndt B 1987 *Ramanujan's Notebooks*part II (Berlin: Springer)

Dougall J 1907 *Proc. Edinburgh Math. Soc.* **25** 114–32

Dunnington G W 1955 *Carl Friedrich Gauss: Titan of Science* (New York: Hafner)

Erdélyi A, Wilhelm M, Fritz O and Tracomi F G 1953 *Higher Transcendental Functions* vols I, II (New York: McGraw-Hill)

Gauss C F 1812 Disquisitiones generales circa serien infinitam *Thesis* Gottingen; pub. 1866 Ges. Werke, Gottingen **II** 437

Gauss C F 1812 Disquisitiones generales circa serien infinitam *Thesis* Gottingen; pub. 1866 Ges. Werke, Gottingen **III**, 123, 207, 446.

Gasper G and Rahmann M 1991 *Basic Hypergeometric Series* (Cambridge: Cambridge University Press). Also ed Gasper G and Rahmann M 1990 *Encyclopedia of Mathematics and its Applications* vol 35 (Cambridge: Cambridge University Press)

Gleick J 1993 *Genius: The Life and Science of Richard Feynman* (New York: Vintage)

Hardy G H 1940 *Ramanujan: Twelve Lectures Inspired by His Life and Work* (Cambridge: Cambridge University Press)

Hardy G H, Seshu Iyer P V and Ramachandra Rao B 1927 *Collected Papers of Srinivasa Ramanujan* ed G H Hardy (Cambridge: Cambridge University Press)

Hoffman P 1998 *The Man Who Loved Only Numbers: The Story of Paul Erdos and the Search for Mathematical Truth* (New York: Hachette)

Kac C M 1985 *Enigmas of Chance: An autobiography* (New York: Harper Collins)
Kanigel R 1991 *The Man Who Knew Infinity* (New York: Washington Square Press)
Musielak D E 2015 *Prime Mystery: The Life and Mathematics of Sophie Germain* (New York: Authorhouse)
Ono K 2016 *My Search For Ramanujan: How I Learned to Count* (New York: Springer)
Ramanujan S 1911 *J. Indian Math. Soc.* **III** 219–34
Singh S 1998 *Fermat's Last Theorem* (New York: Anchor Books)
Slater L J 1966 *Generalized Hypergeometric Functions* (Cambridge: Cambridge University Press)
Srinivasa Rao K 1978 *J. Phys.* 11A L69
Srinivasa Rao K 2004 *Srinivasa Ramanujan : A Mathematical Genius* (Chennai: EastWest Books)
Srinivasa Rao K and Vanden Berghe G 2004 *Historia Sci.* **13** 123
Srinivasa Rao K and Vanden Berghe G 2006 *Ganita Bharati* **28** 7
Wiles A J 1995 *Ann. Math.* **141** 443–551
Young D A B 1994 *Not. Rec. R. Soc. (London)* **48** 107

www.ingramcontent.com/pod-product-compliance
Ingram Content Group UK Ltd.
Pitfield, Milton Keynes, MK11 3LW, UK
UKHW051855100926
8556IPUK00005B/631

9 780750 31494